GUIDE TO
BUYING YOUR
FIRST HOME

Century 21® Editors
and Patrick Hogan

**Real Estate
Education Company**®
a division of Dearborn Financial Publishing, Inc.

This publication is designed to provide accurate and authoritative information in regard to the subject matter covered. It is sold with the understanding that the publisher is not engaged in rendering legal, accounting or other professional service. If legal advice or other expert assistance is required, the services of a competent professional person should be sought.

Executive Editor: Cynthia A. Zigmund
Managing Editor: Jack Kiburz
Interior Design: Lucy Jenkins
Cover Design: ST & Associates
Typesetting: Elizabeth Pitts

Printed in the United States of America

97 98 99 10 9 8 7 6 5 4 3 2 1

Library of Congress Cataloging-in-Publication Data

CENTURY 21® guide to buying your first home / Century 21 editors and
  Patrick Hogan
      p. cm.
    Includes index.
    ISBN 0-7931-2424-7
    1. House buying   2. Real estate business.   I. Hogan, Patrick.
  II. Century 21 (Firm)
  HD1379.C453   1997
  643'.12–dc21   96-52848
                                                              CIP

Real Estate Education Company books are available at special quantity discounts to use as premiums and sales promotions, or for use in corporate training programs. For more information, please call the Special Sales Manager at 800-621-9621, ext. 4384, or write to Dearborn Financial Publishing, Inc., 155 N. Wacker Drive, Chicago, IL 60606-1719.

CONTENTS

Find This Book Useful for Your Real Estate Needs?

Discover all the bestselling CENTURY 21® Guides:

CENTURY 21® Guide to Buying Your Home

CENTURY 21® Guide to Choosing Your Mortgage

CENTURY 21® Guide to Inspecting Your Home

CENTURY 21® Guide to Selling Your Home

CENTURY 21® Guide to a Stress-Free Move

CENTURY 21® Guide to Remodeling Your Home

The American Dream is alive and well.

Home ownership in the United States is at an all-time high. Since 1992, nearly half of all homebuyers have been first-timers, like yourself. Surveys also have shown what we already knew—that would-be homeowners are willing to make sacrifices to get in their own home—whether it's living farther from work, giving up a second car, or canceling vacation plans.

Buying your first home is a big deal. You're treading through unfamiliar territory that has a language all its own. The *CENTURY 21® Guide to Buying Your First Home* is your road map. Perhaps you will make sacrifices to reach your goal. If so, they will be worthwhile. With the advice of this book, you will invest your money soundly and wisely. And 50 money-saving tips will keep you from squandering a penny of your hard-earned cash.

This is your guide, from tentative visits to open houses to signing the papers. You will learn how to work with the professionals you meet along the way. You will understand what you can afford to spend on the home and how to get the most for it. You will know how to read the signposts of neighborhood cycles and the hallmarks of sound house design. You will confidently navigate through the intricacies of the mortgage process.

The biggest obstacle for most people buying their first home is the down payment. You'll learn about mortgages that require down payments of just 3 percent or less and about creative ideas for getting in the door. With this guide, you are sure to make a knowledgeable buying decision.

You can expect this same level of quality when you visit a CENTURY 21® office. Our brokers and associates in your community stand ready to help you put these strategies into action.

All of us at CENTURY 21®–*your* best source for homeowner information and services–wish you the very best in the exciting journey to owning your first home!

CHAPTER 1

Deciding to Buy

Fans of *The Honeymooners* recall Ralph Kramden, fist thumping against his massive chest, proclaiming, "I'm king of this castle, Alice!" But he was forgetting the real king of his Brooklyn walk-up—the landlord. You don't have to own to call your place home. Home is more a state of mind. Yet when your name is on the deed, you have legitimate claim to the scepter. You rule.

If you've picked up this book, you're already giving serious thought to buying your first home. It's a big deal, you know, and you set out on a complex journey, a maze of stumbling blocks and choices. Why bother? Wouldn't it be easier to sign that lease just one more time?

For most people, buying a home is the single biggest investment of a lifetime. It's quite likely that you'll clear out most of your savings to scrape up a down payment. Make this jump and you've started on your nest egg. "What! I just spent it," you say. True, but you will be building *equity*, the difference between the market value of the house and what you owe. It's almost like a tax-free savings account. Maybe you may decide to draw on your equity to give your children a college education. Decades from now, your home could fund your retirement. But now it is a jump. You must consider homebuying as a business investment and handle the process in a businesslike manner.

But buying a home is not strictly a business decision. It's nothing like building a stock portfolio. Emotions run high every step of the way. The home plays host to your most heartfelt memories. On one buyer's final walk-through, the homeseller sat brokenhearted in a half-empty living room, boxes stacked along the wall. Wiping tears from her cheek, she sobbed, "We had our babies here." Yet just a couple weeks ago she was punching numbers at her keyboard, satisfied that profits from the offer they had just accepted would be enough to buy a bigger home in a better school district. Buying a home is both a business decision and a lifestyle choice.

Risks and Rewards

Let's first forget the matters of lifestyle for a moment and take a cold look at the numbers. Buying a home is a real estate investment. Should you be adding real estate to your portfolio? Like any investment, consider the risks and the rewards. Benjamin Franklin believed in confronting a decision with a list of pluses and minuses. Later, we'll take an in-depth look at some of the issues below. For now, let's get the list going with the downside:

- Unless you can qualify for a low down payment program, you will probably have to put up a substantial amount of money. Circumstances may force you to sell at a loss. Could you afford to lose that investment?
- Real estate is an illiquid investment. Unlike your mutual funds, you can't cash in by calling an 800 number available 24 hours a day.
- Investing a large amount of money entails an opportunity cost. You may not be able to take advantage of Uncle Joe's can't-miss stock tip. (Depending on your experience with sure things, or Uncle Joe, you may choose to move this to your plus list.)

- Economic conditions could change. For example, a major employer located nearby may leave the area or close shop. No longer will new employees shop your neighborhood for homes. Displaced workers will put their homes up for sale, flooding the market. Worst of all, the employer could be yours.
- Zoning changes could impact your property. A colossal rental housing development could take over the luscious prairie adjacent to your property line. Not good for your value.

Despite the inherent risk, real estate historically has proven to be an excellent investment. Here's what you can look forward to on the reward side of the equation:

- Property taxes and mortgage interest can be deducted from your state and federal income taxes.
- You'll be building equity. The lion's share of your mortgage payment goes to interest, with the remainder applied to your principal balance. Initially this is a miniscule amount, but it can become significant over time.
- Chances are good your value will appreciate. Real estate agents like to quote Will Rogers on land: "They ain't makin' any more of it." If all goes well, this is how you'll build equity, even if you own for less than ten years.
- You can tap into your equity by refinancing for a larger amount of money or, where available, by getting a home equity loan. These funds will not be taxable and in most cases are tax deductible. (Here is some relief from the illiquidity of real estate.)
- Real estate is a good hedge against inflation. Nothing is guaranteed, but in times of growth, real estate appreciation often outpaces the current inflation rate.

Market conditions will dictate the cost of renting and the value of homes in your area. In many parts of the country you'll find that rents in desirable areas close to the city center and all its amenities exceed the cost of a mortgage payment for an average home in outlying areas, or at least the typical starter. More likely than not though, your monthly housing cost is going to jump when you go from renting to buying. Your monthly payment will be comprised of not only the mortgage but also homeowners insurance and property taxes. However, you will benefit from tax deductions not available to renters. On the other hand, maintenance costs can add a real wild card to the equation. Your costs will vary widely depending on the age and condition of the property. But any homeowner will tell you, "It will always be something." As a renter you may wait a while for a repair to be made, but you shouldn't have to pay for it. When you're king of the castle you pay when the turret needs tuck-pointing.

☞ **Money-$aving Tip #1** *When comparing your rental costs with projected ownership costs, don't forget to factor in maintenance expenses.*

The short of it is you'll be tying up a chunk of money when you buy a home. Could you do better with another investment? Even if you base your homebuying decision on this single factor, the answer is not simple. You can make educated guesses about such economic factors as interest rates, rate of inflation, and stock market and real estate trends. Still, they are only guesses. And then there is so much more to add to the equation.

Real Estate Is Leveraged

Buying real estate is a leveraged investment. When you buy a home, anywhere from 80 to 95 percent of the funds will probably come from your mortgage lender. You are *leveraging* a relatively small down payment (small not in dollars, but as a percentage of value). The lower the down payment the greater your risk. But the leverage makes your funds grow fast. Let's look at an example.

Suppose you buy a house for $150,000 with a down payment of $15,000. In the first year the house increases in value by 5 percent, about the same amount as your money market interest rate. Your home is now worth $157,500. In a one-year period you've made $7,500 from your investment of $15,000. That's 50 percent, not bad! However, keep in mind that there are still costs related to the sale.

Income Tax Deductions

During the 1996 political campaigns, candidates and policy-makers debated the possibility of a flat tax, set rates, and no deductions. Common to nearly all proposals was one exception—the deduction of mortgage interest for a primary residence. Indeed this deduction is treasured among Americans. If real estate and construction lobbies were ever to lapse in their efforts to keep the deduction intact amid tax policy changes, then grassroots middle America surely would pick up the slack.

To take advantage of this benefit, you must itemize your deductions. The rules are that you occupy your residence, and your house must be the security for the loan. The IRS uses the term *acquisition debt* to describe debt secured by the residence and incurred in acquiring, constructing, or substantially improving a principal or second residence. Under current IRS guidelines, your upper limit in deducting interest is acquisition debt of one million dollars, which should cover most of us. Guidelines change, so it is always best to consult a professional tax adviser. In addition, if you refinance for a larger amount than the original debt ("take cash out") or take out a home equity loan, you can deduct up to $100,000 of interest. Owners of condominiums, cooperative units, and town houses also are eligible for this deduction and, under certain conditions, you can even swing it for a houseboat or mobile home.

When you own a home, you'll pay property taxes to fund your schools, parks, police and fire protection, and other municipal services. The money is an extra expense of home ownership, but another deduction softens the blow. Property taxes for all real estate, not just a principal or second residence, are fully deductible. Owners of condominiums or cooperatives can deduct their prorated share of the full amount the homeowners association pays.

Other tax benefits of home ownership will be discussed in Chapter 14. For the sake of comparisons with the cost of renting, the key tax deduction to consider in the home ownership cost tally are the mortgage interest and property tax deductions. You'll take these every year.

Financial Stability

Owning a home is a first step toward financial stability.

In order to qualify for a mortgage, you'll have to demonstrate at least a decent credit history. You probably use some types of credit already. A lapse or two in making your payments on your credit cards is forgivable. At worst you'll be hit with extra financing charges or a penalty fee. You're in a different league when you take on a mortgage. You'll see the difference when you look at the late fee on your first mortgage statement. It's not small change to miss that deadline.

The word *mortgage* is derived from the French word *mort,* meaning dead, and *gage,* meaning pledge. It does not mean, "You break the pledge, you're dead!" But you are taking on a serious obligation. If you don't perform your end, the security is your house. The lender can take it away.

Where's the stability in all this? You *will* pay the mortgage. (Only a small fraction of mortgages go into default, and even fewer suffer foreclosure.) You'll build credit history and will find it easier to borrow in the future. Creditors will respect your ability to take on a long-term, substantial debt and pay it off.

Paying off the mortgage is a forced savings plan. The IRS lets you postpone taxes as long as you buy a home of equal or greater value within two years after selling a home. You pay no taxes as your equity grows. When you sell the home you can continue to avoid taxes with the rollover provision. People tap into home equity for critical expenditures, like starting a business or their children's college tuition.

You also gain greater control over your housing and its costs. You don't have to worry about substantial increases in rent or a landlord converting your building to condominiums and asking you to buy or leave. This is not to say your housing costs will be fixed. One of the reasons landlords raise the rent is property tax hikes, which will directly hit the homeowners' housing costs. And you also will have maintenance expenses.

Joining a Community

When you buy a house, you will be joining a community. Renters, too, are part of their communities. But owners are firmly rooted. As a homeowner you will have a more prominent voice in local matters. For the first time, you will be able to say, "Gosh darn it, I pay taxes in this town, and I don't want . . . ." Community organizations often become formidable political forces.

When you choose your house and location, you also will be choosing your neighbors, schools, parks, churches and synagogues, and merchants. Casual encounters, like a conversation over a backyard fence or chattering with other parents at a Little League game, make for happy days. In the best of circumstances, neighbors soon become friends, sharing life's ups and downs together.

Of course, joining a community can just as easily be a negative. If the culture doesn't fit, you'll lose out on a basic human need—a sense of belonging. Renters easily pick up stakes and move on. Not so with homeowners. Choose your community carefully. Your range of choices cuts to the heart of the lifestyle aspects of buying a home. It's nothing like buying shares of stock.

Spiritual Comforts

"Home is where the heart is" reads the framed needlepoint at the church bazaar. It's corny, and it's true.

Unlike a renter, you will be able to tinker with your home to your heart's delight. Okay, you like mauve walls in your bedroom, why shouldn't you be able to have them? How many leases prohibit "holes in walls"? So you deliberate on hanging a favorite painting and call your landlord to see if it's alright. When you own, only your spouse will restrain your decorating whims.

If you enjoy weekend projects, you'll have no shortage as a homeowner. You're sure to become friendly with your local helpful hardware person. On the other hand, if the sight of a hammer makes your heart pound, you might lighten your maintenance responsibilities with a condominium or town house.

Owning a home will give you a sense of security. You won't have to worry about a call from your landlord with news that Uncle Louie from Detroit is moving into town and wants your apartment.

For many people, however, such stability is not so desirable. You may want the flexibility to take a new job in another city. Or maybe you expect a transfer from your employer. You may want to be able to move to the other side of town just because you've fallen in love with somebody out that way. Traditionally, marriage, home, and kids have all been woven into a fabric called the American Dream. Lifestyles are changing. People are marrying later or not at all. More and more single people are homebuyers. If you're single and do see marriage in your future, should you buy? It may be wise to do so, but plan for the possibility of selling if you do get married. Your spouse may have different ideas about a home or just want to start out in a place that's "ours," not "yours."

There's no arguing with the flexibility renters enjoy. That alone is often important enough to offset all the advantages of ownership. However, owning a home isn't a lifelong commitment. Few people stay long enough to pay off the mortgage of their first home. You can always sell. The problem is you may have to do so at an inopportune time. Also, if you own for a short time, say just a couple of years, any appreciation you've realized is likely to be eaten up in selling expenses. You still get the tax advantages. Many young homebuyers have come to the table because they're getting killed by taxes. If you anticipate a short term of ownership, be sure to tell your real estate agent. You'll want to give special attention to buying a marketable property. And you may want to hold off on the mauve walls.

When to Buy

You've probably been thinking about buying a house for some time. The biggest challenge facing first-time homebuyers is scraping up enough money for a down payment. Perhaps you are already building your war chest. When should you make the jump and buy? The simple answer to that question is when you're ready. But that begs the question, you say, when will I be ready? The coming chapters will help you determine that, at least in terms of your financial situation.

The point here is that you should look at yourself to determine the right time. Don't fix on the market. It is natural to try to time your buy with the market, just as you might in making a stock purchase. Real estate is cyclical, however. You will probably own your home for five years or more. You may quite likely own a residence of one sort or another for the rest of your life, or at least into your retirement. During that period of time, you are going to experience ups and downs in real estate values. To a certain extent, when you jump into the market doesn't matter all that much.

It's nice to be shopping in a buyer's market. Real estate follows the law of supply and demand. The more houses that are on the market, the better the deal you'll be able to negotiate. Real estate is local as well. National trends can be particularly misleading. The government issues statistics by region, but that is not much help either. Even within a town, market conditions will vary. Homes in a popular neighborhood with great schools may sell in days, while the average house in town is on the market at least three months before a sale. The situation of the seller is a bigger determinant as to what kind of deal you will get. For instance, if the house is owned by a couple in divorce court, they may be quite anxious to move the property so they can divvy up the cash.

A "good time to buy" is also characterized by low interest rates. Nearly all first-time buyers will take out a mortgage to finance their purchase. Small swings in interest rates can make a big difference in the effective cost of the house. For instance, if you take out a mortgage of $125,000, the difference between an interest rate of 8 percent and 9 percent will be about $89 every month. That may not seem like much, but multiplied by the number of months you'll own, it certainly makes a difference. And it may even make the difference on whether or not you qualify for the mortgage. Buyers like low interest rates. (And so do sellers, because if funds are readily available they'll get more for the house.) But interest rates too are cyclical. If you buy at a high point, it is quite possible that during the term of your ownership, you'll be able to refinance at a lower rate. You also might gamble that rates will drop by choosing an adjustable-rate mortgage.

The first home purchase is usually a stretch. You'll receive different advice. Some might say buy the best home you can afford and put as little down as possible. More conservative advisers will tell you to wait until you have enough to put 20 percent down; anything less is dangerous. In a rapidly appreciating market, the idea of saving for a home can be as elusive as a carrot on a stick. You squeeze every penny and still fall short because houses are appreciating faster than you can save. Nevertheless, you might find the same phenomenon in certain hot neighborhoods. In this situation it does make sense to act sooner rather than later. However, don't let a sense of urgency force you into a hasty decision. Remember, the home purchase is an illiquid investment. Think long term. Buy a home when you're ready financially and psychologically.

How Your Agent Can Help

Real estate agents will help you on the investment side of your home purchase. They will run down your costs of purchasing and ownership and inform you about trends in your area. Agents have access to a database of all the houses in the marketplace. You don't have to be satisfied with vague generalities. Ask the agent for specific information, such as: What's the average length of time a house is on the market before the sale? What is the average ratio between selling price and asking price?

Agents also can give you qualitative information about neighborhoods. For instance, they'll know what areas are undergoing a lot of renovation. They may know about construction projects or zoning changes that will affect future value. They'll be able to point you to areas full of young families, good schools, or recreational opportunities that fit your interests. However, real estate agents may be hesitant to discuss social characteristics of neighbor-hoods. To discourage discrimination and racial steering, agents are required by law to show you homes in any neighborhood you can afford and are forbidden from making any comments about race.

Much of what agents can do for you depends on who they represent in the transaction. You'll have some choices, as discussed in Chapter 6.

Commonly Asked Questions

Q. We're not quite ready to buy yet, but will be next spring. What should we be doing to get ready?

A. You'll have lots of decisions to make when buying a home, and you may have to act quickly. It is prudent to start laying the groundwork early. First, you should get your finances in order. This is the subject of the next chapter. Second, take some time to scout out neighborhoods. Such explorations can be a fun summertime activity. You won't have the stress and pressures of actual house hunting. Instead, you'll be keenly observant tourists. A great way to get a feel for a neighborhood is to get out your bikes and ride around. Or take a long walk. You'll see all kinds of things that you would miss driving through. Keep your eye out for community events. Attend a service at a local church or synagogue. Shop the garage sales and strike up conversations with the people you meet. Is this a community you can see yourself in?

If you feel that you've zeroed in on a neighborhood, you may want to stop in a few real estate offices in the area. Many people do everything they can to avoid a sales pitch, but you shouldn't feel any pressure. Explain your situation. Most agents will be quite happy to help you with some general market information in hopes that you'll stop back again when you're ready to start looking. Talk to a few and find somebody you're comfortable with.

Q. Is it okay to start looking before we're actually ready to buy?

A. A bit of window shopping never hurts, particularly when you're anticipating a purchase as big as a house. However, be sensitive to the agent's situation. Agents, of course, work on commission and are trained to use their time as efficiently as possible. Agents will lose interest quickly if they show you around a few places and sense that you aren't serious. Best to be up front with agents right off. Give them a general idea of the types of homes you're looking for and the price range. Most real estate offices will enter you into their leads management database and contact you when properties that meet your criteria come on the market.

Your real purpose is to get a feel for the market and what your money can buy. You don't need to look at too many homes to accomplish that. It's not an effective use of your time either to spend too much time looking at houses you're not prepared to buy. And if you do fall in love with one, it will be frustrating that you can't start negotiating. A good strategy is to attend open houses. Not only can you shop houses, but you can shop agents at the same time.

Q. What happens at an open house?

A. Though open houses might bring lots of visitors to a house, they're usually not too effective at selling houses. For one, people that attend may not be buyers. For example, open houses attract homeowners in an area curious about what the place is like inside or looking for a relative feel for what their homes are worth. Open houses also attract people who are just starting a search but are not yet ready to buy. Like you, perhaps. For agents, open houses are good for generating leads of prospective buyers. You don't have to fear that the agent will give you the hard sell on the house. More likely, the agent will want to learn more about you—like what you're looking for, how long you've been looking, etc. The agent will want to qualify you as a lead. At the same time, you can assess the agent as one you may want to work with. Attending open houses is an especially good activity in the early stages of your house hunting.

CHAPTER 2

Your Financial Self-Assessment

According to a *Wall Street Journal* survey a few years ago, more than two-thirds of Americans live from paycheck to paycheck, an astonishing figure. Americans are consumers, not savers. Our national savings rate has hovered at less than 2 percent for the past decade, even in the heydays of the 1980s. It is therefore no wonder that scraping up enough money for the down payment is the biggest challenge for most would-be homebuyers.

Before you take on the financial obligation of a mortgage, you should take a close look at where you are financially. You can be sure that your lender will. Lending six figures on a 30-year term isn't something even large institutions take lightly. To get some idea of the scrutiny you can expect, ask your bank for a mortgage application.

You will receive a worksheet for filling out your assets (what you have) and your liabilities (what you owe). Subtract what you owe from what you have and you get what you're worth. The bank also will ask you to sign a release permitting it to dig into your credit history. Before you fill out an application, you should know your net worth and what credit reports are saying about you.

The mortgage payment for your home will most likely exceed your current rent. So if you're like most folks and have little left over at the end of the month, it's time to review your spending patterns and see where you can cut back. Your lender will examine your income and your debt levels to determine your ability to make the monthly payments on the mortgage. Before you start looking, you need to make your own determination of how much house you can afford. So you need to know where your money is going. What the bank considers to be the top mortgage you can afford may be more than what you're comfortable with. You don't want to cut back too much, denying yourself life's pleasures. If you enjoy frequent ski vacations, why not buy a smaller house and continue to hit the slopes. Sure beats languishing around a luxurious house, dreaming of past pleasures.

☞ **Money-$aving Tip #2** *Even while you sacrifice to afford a first home, set some fun money aside. You don't want to be "house-poor."*

Determining Your Net Worth

You might be surprised just how much you're worth. Your assets include everything you have, not just your money in the bank. Imagine holding a "garage sale" in which you open the doors of your home for an "everything goes" extravaganza, and everything does go. All the little things do add up. Of course, such personal property is not going to help with your down payment unless you do indeed hold this wildly successful sale. However, it will factor into the lender's evaluation of your creditworthiness. On the other hand, if credit card debt is powering your lifestyle, you might find that in the balance you don't have quite as much as you thought. Here's what to factor in on both sides of the equation:

Assets

- *Cash.* Includes all the money in your savings and money market accounts

- *Securities.* Include stocks, bonds, mutual funds, treasury securities, etc., that can be converted into cash fairly readily.
- *401(k)s, 403(b)s, or IRAs.* These assets cannot be converted into cash without paying penalties and accepting less than face value. You don't want to use this money before its time, but it does nevertheless factor into the balance sheet.
- *Automobiles.* The full value of your car if you were to sell it on the used car market.
- *Other personal property.* Include everything—furniture, collectibles, computer, stereo, TVs, art, clothing, tools, and more.
- *Insurance policies.* Include cash value of life insurance policies.
- *Real estate owned.* You probably don't have any if you're reading this book, but you might. Include any real property you own in whole or part, such as a share in a family vacation home.

Liabilities

- *Loans owed.* Include automobile or boat loans, furniture or appliance loans, student loans, and all installment debt.
- *Credit card balances.* Include all card balances no matter how small. (Even if you pay off the credit card every month, banks factor whatever the current balance is at the time of underwriting.)
- *Income taxes payable.* In case you fell behind with the IRS.
- *Other bills due.* Include unpaid doctor or hospital bills, past due child support or alimony, or whatever else you owe.
- *Real estate mortgages.* Probably none . . . yet.

Run down the list on a worksheet and tally up the numbers.

Assets – Liabilities = Net worth

Where Does All the Money Go?

A great feature of the check-writing software programs that are becoming ever more popular is that they make it easy to keep detailed records on how you are spending money. If you're using Intuit's Quicken, Microsoft Money, or similar software, be sure to set up categories for your expenses and use them. With the click of a button you'll be able to access information that will be useful as you prepare to buy a home. In fact, if you have a computer and your home purchase is still six months or so away, consider purchasing one of these programs.

While not as detailed or accurate as the records generated by computer, paper and pen can still do the job for you. Get a small notebook that you can carry with you at all times. For a period of at least two weeks, but preferably a month, record every purchase you make. No exceptions. If you buy a newspaper or a candy bar on your way to work, write it down. Your goal is to get a feel for your monthly living expenses so you can fill out the worksheet in Figure 2.1. For your bill-paying cycles, your checkbook register can serve as the record. Refer to credit card statements to pin down your spending habits on bigger ticket items. Some expenses will be difficult to allocate as a monthly expenditure. It would be nice to take a vacation monthly, but you probably take just one or two a year and a few weekend getaways. For these items, make your best guess at what you spend annually and divide by 12.

Save Money Too

Notice that savings is on the list. The inclination is to save what's left over after you've paid off all your bills. The problem with that kind of thinking is that there won't be much left over. Personal finance experts will tell you that the best way to save money is to "pay yourself first." Make it your goal to save at least 10 percent of your gross income. Stick to that plan and you will always be living within your means as well as building a nest egg for life's big expenditures, like college tuition and, yes, buying a home.

FIGURE 2.1 Monthly Spending

Rent	_____
Food	_____
Housing	_____
Utilities	_____
Telephone (include cellular & pagers)	_____
Insurance (auto, health, life)	_____
Medical & prescriptions	_____
Car payments	_____
Gas and car upkeep	_____
Commuting to work	_____
Kids' costs (child care, tuition, toys)	_____
Alimony & child support	_____
Entertainment	_____
Dining out	_____
Vacation(s)	_____
Clubs and associations	_____
Gifts and charity	_____
Savings	_____
Everything else (Add 5% or so)	_____
TOTAL	_____

Another obstacle to saving is the temptation to dip into the nest egg. Paying your children's college tuition may be a reason to dig in, exotic vacations or a new car aren't. Instead, you should set up a separate account for such one-time expenses, which you know will come up. Another rule that will develop a savings discipline is to make it easy to save money and hard to spend it. The best example here is the 401(k) program offered by many employers. You designate a percentage of your salary to be deducted before taxes and deposited into your 401(k) account. Many employers also will match all or a percentage of your deposits. Regulations governing these accounts make it particularly hard to spend your money. You will pay substantial penalty charges for early withdrawal. A less extreme example is to use an automatic investment program to invest in a mutual fund account. You'll still be able to withdraw the money fairly easily in a pinch, but it won't be quite as tempting as a savings or money market account at the bank where you do your checking.

☞ **Money-$aving Tip #3** *Though depositing money into tax-deferred retirement accounts like 401(k)s, 403(b)s, and IRAs is an excellent savings strategy, you may want to limit your contributions until you have enough money to make a down payment on your first home, which also has its own tax advantages.*

Your Credit History

After the lender receives your mortgage application, the underwriting process begins. Underwriting is the lender's analysis of its risk in giving you the loan. Key in this analysis is the lender's assessment of your ability to repay the loan. The best way to demonstrate your creditworthiness is a track record of timely payments.

You know best where you stand. Now is the time to be on your best behavior. Don't make any distinctions between important bills and ones you can let slide. You never know which creditors will be reporting late payments and which won't. There isn't always much logic to it. For instance, bank cards and some oil and gas companies will report regularly, whereas car dealers may report only if an account is long past due or the repo man has been called. Imagine ignoring your car loan for three months without a blemish on your record, only to be tripped up when you let your gasoline card bill slide a couple of weeks. It could happen.

Benjamin Franklin's Poor Richard advised, "Neither a borrower nor lender be." My, how times have changed. But if you're a person who has taken that advice to heart and paid with cash, then you might be in trouble when it comes to getting a mortgage. Unfair as it may be, no credit history is not much better than a bad one. You should apply for credit cards and start using them. Department store cards are a good place to start. You also should look into the major consumer credit cards, though you may need a bit of a credit history to get one.

At the other extreme, if you are the type of person who says yes to every offer of credit that comes through the mail, you also may be in for trouble. It doesn't matter if you pay all your bills on time. The lender may view all your accounts as a tremendous potential to get yourself into trouble, which they certainly are, and that makes you a bad risk. For the same reason, you don't want to apply for a lot of new credit when you anticipate that you'll soon be trying to get a mortgage. Your credit report will list creditors and others who have checked your file in the past six months.

Your Credit Report

Just as important as your credit history is what credit reports say your history is. This distinction is important. The credit reporting industry is a huge database business whose customers are the creditors. Unfortunately, history has shown credit reporting bureaus to be less than responsive to the consumers on whom they keep data.

Your information may be commingled with somebody who has a similar name. A former spouse may come back to haunt you. If you're a junior, make sure your parents are paying their bills, particularly if you are still living with them! Account information may be incomplete. A past delinquency may show up without any indication that it was paid off.

☞ **Money-$aving Tip #4** *The Fair Credit Reporting Act (FCRA) prohibits credit bureaus from keeping most negative account information on your report for more than seven years, or bankruptcies for more than ten.*

Obviously, you want to see your report before the underwriter does. The "big three" credit reporting agencies are TRW, Trans Union, and Equifax. You can contact a credit reporting agency directly to get a copy of your report, usually for under $10. If you have been denied credit, the report is free. TRW will provide one free report annually whether or not you've been denied credit.

TRW National Consumer Assistance
P. O. Box 2104
Allen, TX 75013
800-682-7654

Trans Union
Consumer Disclosure Center
P. O. Box 390
Springfield, PA 19064-0390
610-690-4909

Equifax
Office of Consumer Affairs
P. O. Box 105873
Atlanta, GA 30348
800-685-1111

Correcting Errors in Your Report

If you find errors in your report, you should contact the credit bureau in writing. A dispute resolution form may have been sent with your report. If so, fill it out and return it with your letter. You don't want the company to use that as an excuse for delaying the correction. If possible, include a letter from the creditor as well to support your claim. Send everything certified mail and request a return receipt. Correcting the error may take persistence. Be meticulous in your record-keeping. Maintain a detailed log of telephone conversations, including names of people you talked with and what was discussed. Keep copies of all correspondence. The credit bureau should send you a new report once the problem is corrected. Of course, keep this for your files as well.

Not all creditors report to credit bureaus. Therefore, positive information on your payment history may be missing. If you feel that information is critical to your getting credit, you can take steps to have it added. You will have to contact the credit bureau in writing, stating the name, address, and phone number of the creditor and your account number. Credit bureaus are not required to add this information, however. They may charge a fee for it. They'll also verify it, so don't trump anything up.

☞ **Money-$aving Tip #5** *An excellent resource for working your way through credit conundrums is* The Credit Repair Kit *(second edition) by John Ventura (Dearborn Financial Publishing, Inc., 1996).*

When the Problem Is Accuracy

If you've had a problem or two in the past, it probably won't come as a surprise to find negative information on the report. Unless the problem is more than seven years old, you'll have to live with it. The FCRA does give you the right to insert a written statement into the credit company's file explaining the reason for the problem. However, there is no guarantee that creditors will see that statement, and, in fact, chances are they won't. Nevertheless, if extenuating issues have caused a problem in an otherwise solid record, it's a good idea to send in a brief statement explaining how the problem affected your finances and what you are doing to get back on track. Keep a copy of the statement and submit it with your mortgage statement. Our economy is such that bad things do happen to good people. Health care costs are a perfect example. A medical problem deemed a preexisting condition could set you back significantly. You can hope that a lender will be forgiving. If underwriters see a credit report problem, they may very well contact you for a written statement anyway. By anticipating and disclosing the problem up front, however, you are demonstrating a level of responsibility that may help your case.

You might also try negotiating with the creditor to get the information removed. This is a long shot, but may be worth a try. You'll have the best chance if you have overcome the problem and have rebuilt your business relationship. It's always best to handle these negotiations in person when possible.

☞ **Money-$aving Tip #6** *Do not use credit repair or fix-it companies to clear your credit. They may charge exorbitant fees to do nothing more than what you could do yourself. Or worse, they may resort to illegal techniques, putting you at criminal risk as well. These companies also may pose as financial counseling firms.*

Scraping Up the Down Payment

There's no instant solution here. Just tighten your belt. It may take time. In the early 1990s frugal living spawned a whole industry, or at least a line of books and newsletters. No doubt this was a reaction to the spendthrift 1980s.

Are you willing to sacrifice to buy a home? Study your spending worksheet. What can you cut out? You should see some obvious choices. Try doing without the season's latest fashions. Dining out can really eat up your funds. Next time you need a night off from the kitchen, why not get a cheap bottle of wine and order a pizza instead of going to that Italian restaurant. You bachelors, try using that cookbook your mother bought you.

If you're a two-car couple, perhaps you could sell one. If you're thinking of buying one, don't. America is in love with its cars, so this strategy is no doubt radical. But if you have alternatives, such as carpooling or taking public transportation to work, give it some consideration. You'll sacrifice convenience, but what better way to come up with a few thousand dollars quickly. Any other large, salable asset is also fair game. Your boat. Your Super Bowl tickets. Be creative.

If you're lucky enough to have generous family members who will help you with the down payment, you may want to start moving the money into your account well ahead of time, particularly if it's in the form of a friendly loan. Lenders will notice any large influxes into the account during your evaluation period. If significant funds are coming from a third party, the lender will want a signed gift letter, which is a formal document stating that the funds will not have to be paid back.

How *Your Agent Can Help*

You're the one who must take a cold, hard look at your finances. Your real estate agent can't help there. Once you do, you can get a lot of help from an agent because you'll know where you stand. In later chapters we'll talk about why and how you must share personal financial information with real estate agents. The purpose of this chapter is for you to have a good idea of how much cash you have to put down on the house and what kind of monthly housing costs you're comfortable with. You also need to be certain of what credit reporting agencies are saying about you. With this information in hand, you're ready to work with the real estate agent in mapping out a strategy for your home hunt.

Commonly Asked Questions

Q. I've had credit problems, but that was in the past. Will I be able to get a mortgage?

A. There's no easy answer. Underwriting is about risk assessment. Your record during the past two years is especially important. The lender may determine that you qualify for the loan but that your past problems put you in a high-risk pool, and therefore the interest rate should be higher. It's a judgment call. Therefore, you may want to go to a lender with whom you already do business and have a relationship.

Q. How much money will I need for a down payment?

A. Conventional financing requires a down payment of 20 percent of the purchase price. Fortunately, all kinds of "unconventional" financing are available. (In fact, these are almost the convention for first-timers!). Most lenders offer loans with 10 percent down, provided the borrower purchases private mortgage insurance (PMI), which protects the bank if you default.

PMI is discussed in Chapter 10. More and more home afford-ability programs are cropping up, allowing down payments of just 5 percent or less. You also may be eligible for federally insured loans from the Federal Housing Authority (FHA) or the Department of Veterans Affairs (VA), which require down payments of only a few percentage points.

Q. Can I use my 401(k) or IRA money for a down payment?

A. You can't withdraw money from these funds without incurring substantial penalties. However, some employers permit you to borrow against a 401(k) in certain instances, like purchasing a home. You would then work out a payment schedule with principal and interest going back into your account. You are truly borrowing from yourself because you keep the interest. Although you will be borrowing funds for the purchase of a home, the interest will not be tax deductible because the property won't be attached to the loan as collateral. You also should check with your lender on how they view this in the underwriting process. Of course, it becomes another item on the liability side of your balance sheet.

Congress during the past few years has continued to raise the possibility of allowing IRA withdrawals for a home purchase. As of this writing it is not allowed, but stay tuned.

CHAPTER 3

A Banker's-
Eye View

You've taken a close look at your financial picture. You know how large a mortgage payment you can handle. You've thought about what sacrifices you'll need to make to swing it. These are issues for you to settle. It is your lender, however, who has the final say-so on whether or not you get the loan.

The underwriting process is basically the lender's assessment of risk. Whenever a lender makes a loan, it takes the risk that it will not be repaid. Suppose your friend is caught short at the shopping mall and you fork over $20 to help him out till he gets to his stash at home. In a lightning flash you've probably asked yourself the questions, "Will he pay me back?" and "What if he doesn't?" (The mortgage industry would say you've made a "no-doc" loan—without documentation—but we'll get into that later.) When you walk into a bank or savings and loan off the street and ask for $100,000, which you'll pay back over the next 30 years, the lender asks the very same questions. But they want an answer that includes documentation, and usually lots of it.

In Chapter 10 we'll delve into your various options in mortgages. Here we'll discuss the lending system, what lenders need to know and why, and what size mortgage you will be able to qualify for. With that information in hand, you'll be ready to set your price range and start looking.

The Secondary Mortgage Market

The secondary market is good place to start in looking at the mortgage lending process. When you take out a mortgage it is extremely unlikely that your lender will keep it for the life of the loan. Instead, the lender will sell it. The secondary market is where existing mortgages are bought and sold. The market creates liquidity. The concept is simple: If a lender is doling out mortgages of $100,000 to $200,000 here and there, all to be repaid over an extended period of time, its pool of money would quickly deplete. And when you walked in eager to get into your dream house, the loan officer might say, "Gosh I just lent out my last hundred grand, try the Second National Bank." The secondary market enables the lender to sell your mortgage and lend the proceeds to somebody else. You may not even know your loan has been sold. Your lender may continue servicing the loan, but somebody else actually holds the note.

The largest of the secondary market purchasers is the Federal National Mortgage Association, popularly known as Fannie Mae. Originally founded as a federal agency, Fannie Mae now operates as a private corporation but still maintains ties to the government. Its status as an "agency" of the government boosts its clout in the financial marketplace, enhancing its credit rating and allowing it to borrow funds at a lower rate than most corporations. You may have noticed by now that Uncle Sam is trying to help you afford to buy a home. Fannie Mae was founded in 1938 to purchase Federal Housing Administration (FHA) loans and encourage lenders to make them despite the risks of the depression economy.

Other big players in the secondary market are the Federal Home Loan Mortgage Corporation, or Freddie Mac, and the Government National Mortgage Association, known as Ginnie Mae. Fannie Mae and Freddie Mac purchase conventional, FHA, and VA loans; Ginnie Mae purchases FHA and VA loans. The secondary market includes smaller players as well, but only the big ones get to have cute nicknames.

For this marketplace to work, loans must become a standardized commodity. The secondary market lenders therefore set guidelines as to the type of loans they will buy. Fannie Mae and Freddie Mac have cooperated to develop a series of uniform mortgages, with specific rules on the size and other features. In so doing they have created a class of conventional loans that are called "conforming." These agencies follow national housing policies. By declaring a type of loan as acceptable for purchase, they essentially create new products. For instance, the Community Homebuying Program, an affordability program that will be discussed in detail later, came about because Fannie Mae agreed to purchase these highly leveraged loans. The guidelines of your lender probably will be much the same as Fannie Mae's. Because there are other players, lenders do have some flexibility.

Getting Your Paperwork Organized

Your lender will need lots of information from you. Be prepared for a bit of legwork getting all the documentation together. The whole process will go much smoother if you provide everything that's needed. Don't try to hide anything. The lender will find the information one way or another. In Chapter 2 we discussed some of the personal financial information the loan underwriter will look at; much of it will require some documentation. Here is a rundown of information lenders typically request:

- Your home address and two prior home addresses. (You should have addresses available for the past seven years.)
- Names and addresses of landlord(s) for the past two years.
- Names, addresses, and phone numbers of your current and previous employers. If the company has multiple locations, use the information for the facility with the human resources department.
- IRS W-2 forms and possibly your full federal income tax returns for the past two years.

- Your two most recent pay stubs. It's good to keep pay stubs for 12 months so you can document raises or regular bonuses.
- Locations and account numbers for all checking, savings, money market, and mutual fund accounts. You also should have bank statements for the past two months.
- Information on your IRA and 401(k) accounts.
- Verification of extra income you want to use in qualifying for the loan, such as child support or alimony, disability income, bonuses, pensions, etc. To use child support or alimony as income, you must have been receiving payments for at least 12 months and be able to verify that payments are to continue for the next 36 months. For bonuses to count, you must be able to show that you have received them for two years.
- If relatives or friends are helping you with the down payment, you may need to provide a gift letter assuring the lender that the money will not have to be repaid. (The rules regarding gifts are discussed in Chapter 9.)
- Statements or other documentation of all debts, including credit cards, student loans, car loans, and credit union loans.
- An explanation letter explaining any judgments, bankruptcies or foreclosures, or other credit problems.
- If you're not a U.S. citizen, your Certificate of Resident Alien Status (green card).
- Last, but not least, your loan application fee.

☞ **Money-$aving Tip #7** *If you're self-employed or in a business partnership, the paperwork requirements can be difficult to fulfill. Lenders may relax documentation requirements for entrepreneurs they consider to be good risks. You'll still need your papers in good order.*

Employment and Income

Gone are the days of our social contracts with employers: work hard, be loyal to the company, and you will be rewarded with raises, promotions, and a generous pension when you retire 30 or 40 years down the road. Most of us change jobs, and even careers, more than our parents would ever imagine doing. Five years seems to be a long time on the job today. So what are lenders looking for when they review your employment history?

Stability is still king, but it doesn't necessarily mean the same job. Lenders understand that changing jobs is a natural fact of working life. In fact, showing a consistent pattern of moves to better jobs in your same field is advantageous. However, lenders will be more critical about career changes. You can understand the logic. If you were to lose your job, the longer you've been in a field the easier it will be to get a comparable or better job. Also, career changes usually involve some compromises, maybe starting over at the bottom. The lender also will give some consideration to your field of work. For instance, if you work in a manufacturing industry that is losing jobs to overseas production, your taking computer classes and going into data processing would probably be perceived as a positive. Obviously, working in a growing industry is preferable to one that's shrinking. Whatever makes you feel more secure in your employment will make the lender feel better too.

Part-time or freelance income may or may not count in your lender's calculations. Again, lenders look for a pattern and want to be reasonably sure the income stream will continue. Overtime income probably will not help you in qualification. The problem is that it's hard to prove that it will always be there. An exception might be if you work in a seasonal business that by its nature will offer you regular opportunities for overtime work. You may have to provide a letter from your employer verifying that the overtime will continue.

Self-employment, more and more common these days, still presents real challenges in qualifying for a mortgage. The lender will certainly require two years of income tax returns, perhaps more. You also will need to provide up-to-date financial statements, such as balance sheets and profit and loss statements, and a business credit report. The lender may require that this documentation be audited or prepared by a professional accountant. If your business is a corporation or partnership, the lender will require two years of tax returns and the financial statements. They also will want your K-1 showing your percentage of ownership and income and liabilities. If you work on commission, lenders may require a year-to-date income statement as well as two years of tax returns.

Note that the lender will be looking at the self-employed person's *net* income, after expenses, not the gross. Most entrepreneurs will aggressively look for any possible business expense to keep the tax bill down come April 15. The problem is those deductions count against you when you're dealing with lenders. Certain expenses, like depreciation or a Keogh plan or IRA contribution, will be added back, but most everything else is subtracted from income.

☞ **Money-$aving Tip #8** *Self-employed homebuyers should try to work with a lender they already do business with. The lender might be willing to relax the documentation requirements if you have a solid payment record on loans.*

Debt

The less, the better, as long as you have some. Lenders will measure your income against your mortgage payment and all other monthly debt payments. They will set a ceiling on debt as a percentage of income. You should make a conscious effort to reduce your debt. Lenders report that one of the most common debt obligations that push people over the qualification limits is car loans. In the previous chapter we mentioned the possibility of selling a car to raise money for your down payment. If you're making payments on a new car, it's going to make a real impact on your debt ratios. Explore alternatives. Perhaps you could sell your car, pay off the loan, and still have enough money left over to buy a decent used car.

☞ **Money-$aving Tip #9** *If you have problems with debt, you can get low-cost advice through Consumer Credit Counseling, a nonprofit agency. Look in the yellow pages for an office in your area. Also, many community colleges offer inexpensive courses or seminars on personal finance.*

Credit cards can get people into debt problems in no time. Try not to use them. Pay cash as you go, and you'll realize how much you're spending. It's quite an effect. Counting out your twenty-dollar bills to buy that extra pair of shoes you really don't need feels much different than signing a credit receipt. While you're limiting your new purchases, make an effort to pay off any balances you're carrying. Even if you're not running up balances, the access to credit alone may concern the underwriter. If you have accounts you don't use, close them.

Qualifying Ratios

The most important of the guidelines the loan underwriter will use in reviewing your application are qualifying ratios. The *housing ratio,* sometimes called the front ratio, is your mortgage payment divided by your total gross monthly income. The *debt ratio* is the sum of your mortgage payment and all other monthly debt payments (car loans, student loans, credit cards, installment debt, etc.) divided by your gross monthly income.

☞ **Money-$aving Tip #10** *Interest, dividend income, or payouts from trust funds also will count toward your monthly income.*

Your mortgage payment includes not only principal and interest for the mortgage itself, but also your homeowners insurance and property taxes. An easy way to remember what makes up the payment is the acronym the real estate industry uses to describe it—*PITI* (principal, interest, taxes, and insurance).

Typical ratios that the bank would use are 28 percent for housing debt and 36 percent for total debt. The shorthand way of expressing that is 28/36. Loans with these ratios conform to Fannie Mae guidelines. Lenders are under no legal obligation to use these ratios. Although they are common, don't take them as a standard. One of the first questions you should ask a potential lender is what qualifying ratios they use. Lenders may allow you to exceed these ratios, for instance, in return for a higher rate of interest. Or they might use higher ratios for a down payment of 20 percent or more.

☞ **Money-$aving Tip #11** *Homeowners association fees count as housing debt in determining your ratios.*

Let's look at an example. Bob is making $48,000 and has no other sources of income. His monthly debt payments for his car, student loan, and credit cards total $600.

Dividing his annual income of $48,000 by 12, he determines that his monthly income is $4,000. Assuming a front ratio of 28 percent, the lender would consider the maximum PITI payment he could afford to be $1,120:

$$\$4{,}000 \times .28 = \$1{,}120$$

Assuming a back ratio of 36 percent, the lender would consider the maximum Bob could afford for PITI and all other debts to be $1,440:

$$\$4{,}000 \times .36 = \$1{,}440$$

Either the front ratio or the back ratio will be the limiting factor. In Bob's case, it's the back ratio. Because Bob has a monthly debt of $600, the maximum PITI payment he can qualify for would be $840:

$$\$1{,}440 - \$600 = \$840$$

To determine from that number how large a mortgage he qualifies for, Bob would need to know the cost of insurance for his home and the property taxes. As a rule of thumb, you can guess that about 20 percent of the housing expense will go for insurance and taxes and the remaining 80 percent for principal and interest. Using this guideline, Bob figures he can afford a principal and interest payment of $672:

$$\$840 \times .80 = \$672$$

☞ **Money-$aving Tip #12** *You should have an idea of the going rate for property taxes in a neighborhood and what insurance will run. Only then will you have an accurate idea of the price range you can afford.*

So what size mortgage will require a payment of $672, Bob's upper limit? It depends on the interest rate. Figure 3.1 is a factor table for 15-year or 30-year loans at various interest rates. Use the factor at a given interest rate and loan term to determine the monthly payment in principal and interest for every $1,000 borrowed. Or, as we'll do for Bob, you can work the other way and calculate the mortgage amount for a given principal and interest figure. To do that, locate the factor for a given interest rate and loan term. Then divide your principal and interest amount by the factor. Remember the table shows factors for every $1,000 borrowed. So you'll multiply the result by 1,000 to get the total mortgage amount.

Suppose Bob wants a 30-year mortgage and the going rate is 8 percent. The factor for such a loan is 7.34. His maximum principal and interest payment is $672.

$$\$672 \div 7.34 = 91.55 \times \$1,000 = \$91,500$$

Bob can qualify for a mortgage of approximately $91,500. Here it is worth noting the effect that debt can have on the amount of mortgage you will qualify for. If Bob carried less debt and the housing ratio were the limiting factor, going through these same calculations would yield a maximum loan amount of $122,000.

☞ **Money-$aving Tip #13** *Qualifying ratios for government-insured loans are higher. The current FHA ratios are 29/41, VA are 41/41. (Check with your lender or mortgage broker/banker for the latest ratios.)*

You should make these calculations early on to get a ballpark idea of how expensive a house you can afford. But don't lose sight of your own comfort level with payments. You don't necessarily have to buy the most expensive house you qualify for. However, because for most people buying a first home is a real stretch, chances are you will buy at close to your limit.

You can use the factor table in Figure 3.1 to determine if a home is in your range. Suppose you're looking at a home for $130,000 for which you'll make a down payment of 10 percent, or $13,000. You'll need a mortgage of $117,000. If the going rate is 8 percent, your monthly principal and interest payment would be:

$$\$117 \times 7.34 = \$858$$

FIGURE 3.1 Monthly Payment Factors

Interest Rate	15-Year	30-Year
5.0%	7.91	5.37
5.5	8.17	5.68
6.0	8.44	6.00
6.5	8.71	6.32
7.0	8.99	6.65
7.5	9.27	6.99
8.0	9.56	7.34
8.5	9.85	7.69
9.0	10.14	8.05
9.5	10.44	8.41
10.0	10.75	8.78
10.5	11.05	9.15
11.0	11.37	9.52

Then after asking yourself if you'd feel comfortable paying $860 plus the taxes and insurance, you can run the calculations to see if you would qualify under the lender's ratios. For down payments of less than 20 percent, you may need to purchase private mortgage insurance (PMI), which covers the lender in case you default. You, not the lender, will pay the premium. PMI should be included with your mortgage amount in making qualification calculations, so the figures in the above example might be a bit high. PMI will be discussed in Chapter 10.

☞ **Money-$aving Tip #14** *The Federal Reserve has introduced a software program called "Partners," which it is distributing at no charge to lenders, community groups, government agencies, and consumers. The program requires an IBM-compatible computer running under Windows. Its purpose is to provide low- and moderate-income families with a quick, comprehensive understanding of what they can do to qualify for a mortgage. Ask a lender or contact the community affairs office of one of the Federal Reserve's 12 regional banks for more information.*

Prequalification Programs

Most lenders nowadays are offering some type of prequalification program. Prequalification can mean a few different things, but basically it is a lender's estimate of your borrowing power. Prequalifying can help you in a couple of ways. First, when you visit real estate agents they will want to size you up financially to see if you are able to afford a home. By prequalifying beforehand, you can present the real estate agent with a reliable and trustworthy estimate of your wherewithal, without having to share the details of your finances. Prequalification is not an iron-clad assurance because it depends on the accuracy of the information you provide the lender. The information will be thoroughly checked when you actually apply.

The second benefit of prequalifying is that it strengthens your position in the negotiations. Most purchase agreements will include a financing contingency, stating that your commitment to purchase the property depends on gaining approval for a mortgage at a given interest rate within a given period of time. That's a big "if" for any seller. Some prequalifications, more precisely termed preapprovals, are actual loan commitments, in which case you could waive the contingency altogether. Even if your prequalification is only an estimate, the seller will be more comfortable with your borrowing power and therefore more willing to accept the contingency.

Ask your lender lots of questions about the prequalification program so you have a clear understanding of what you're getting. Many lenders will prequalify you for free, in hopes you'll come back when it's time to get a mortgage. Free prequalification programs are usually estimates only, not actual loan commitments. The lender will do a credit check and you must provide accurate and complete information.

If you want to be preapproved for a mortgage you will probably need to pay a small fee, usually under $100. Be sure to read all the fine print. Your loan commitments might not guarantee the rates. Or the lender may charge a fee to lock in the rates. Oftentimes, the rates for a preapproval program will be slightly higher than the lender's other programs.

A driving force of prequalification programs and preapproval programs is that the mortgage underwriting process has been notoriously slow. As lenders automate the underwriting process through computers and networks, the time line is shrinking. Loans that used to take four or five weeks to underwrite are now approved in hours, even minutes. This is most likely to be an option if your loan is a "slam-dunk"—you're putting 20 percent down, you meet all the guidelines, and your credit report is clean. Here again, be sure of what you're getting in a quick approval loan. It's not uncommon to pay a premium interest rate in return for swift processing.

The House Must Qualify

The focus of this chapter is evaluating, from a lender's perspective, how much house you can afford. However, the lender also will be evaluating your property in the underwriting process. Most likely the lender will hire a certified appraiser (at your expense) to do a full appraisal on the property. In effect, you and the seller have determined the market value of the house. However, because the house is the lender's security on the loan, the lender will want a more formal analysis.

The appraiser collects data about the property, the market, the location, zoning regulations, or any other factors that influence the value of the house. An appraiser takes into consideration such factors as the cost to rebuild a comparable house on the same site less depreciation and the recent sales of houses and land in the same neighborhood, and adds a good dose of professional judgment based on experience.

How *Your Agent Can Help*

Some real estate brokerages also run mortgage broker operations. Or a mortgage company with whom they are affiliated may work out of the same office. They will have all the information on programs in your area, as would any mortgage broker.

But even agents who have no business interest in selling mortgages can be great resources as you explore the market. Agents can help with your calculation of qualifying ratios by telling you about property tax rates in various neighborhoods and what you can expect to pay for homeowners insurance. Agents will be able to explain the mortgage application process and recommend a few lenders. They'll also have the most up-to-date information on prequalification, speedy underwriting, and documentation requirements. Technology is rapidly changing the underwriting process and mortgage shopping in general. An agent working in the field every day will be your best bet for information on the latest trends.

Commonly Asked Questions

Q. Will I still be able to get a mortgage if I can't meet the lender's guidelines?

A. The guidelines described in the example in this chapter, 28 percent for housing debt and 36 percent for total debt, come from Fannie Mae. Individual lender's guidelines may vary.

A loan is not out of the question if you exceed a lender's ratios, but you'll have to be able to make an argument for it. Underwriting mortgage loans is as much an art as a science. Factors like your net worth and credit history also will come into play. The lender will consider mitigating factors. For instance, if you are a couple living on one income but anticipate a second income soon, when your children enter school, the lender might be willing to relax the standards. A lot depends on whether the lender keeps loans in its portfolio or who it sells to in the secondary market.

The government-insured FHA and VA loans, which will be described in detail later, use higher ratios. Lenders also may have programs with more relaxed standards, but you will probably have to pay a higher rate of interest. Later in the book we will look at creative finance alternatives.

Q. What are the common problems that can hurt my chances during underwriting?

A. Verification is what takes time in the lending process. Many problems involve verification of employment, down payments, and secondary income sources, such as child support. Stability is the key here. Lenders do not want to see anything out of the ordinary. For instance, if there's an unusually large deposit into a checking account, your lender may suspect that money for your down payment is coming from an outside source, and is in fact additional debt. It's best if you can keep things the same. Don't apply for new credit cards. Certainly don't take on any large debts, such as a new car loan. If your situation changes, notify the lender. Don't wait to see if they find it out themselves. Even a positive change, like starting a new job with a hefty salary increase, will disturb the lender if caught by surprise. They may choose to start the whole process over again.

Q. What information will the lender ask for in prequalifying?

A. The lender is looking for the same type of information as if you were applying for a loan. Basically, the lender is interested in four basic areas: your income, your debt, your credit history, and your ability to repay the loan. Your job history and length of time in the position is also important. The difference in prequalifying is the level of verification. The lender will rely heavily on easily obtained information such as your credit report. They will not request paper documentation from you, though you will have to supply the same type of information as you would with a loan application.

CHAPTER 4

Who the Players Are

Daunting as buying a home may be, the good news is that you won't have to go it alone. In the course of the transaction, you will be working with professionals who do this kind of thing every day. You should choose these advisers wisely. In the end, it is you who will make the decisions. But as a first-timer, you will need to rely on their expertise. Therefore, in addition to professional competencies, honesty is probably the most important characteristic to look for. You want to be able to trust these people. When you start second-guessing them, you'll also be second-guessing your own decisions.

We'll be discussing how to work with these various specialists throughout the book. You should start building your team of experts before you actually need their help. You may not have much chance to look around when the time comes. Here, we'll give you a quick rundown of who the players are—what they do, when they'll do it, and what kind of credentials to seek. So get your lineup and scorecard ready.

Real Estate Agent

If you have a right-hand person in your home search, it's the real estate agent. First, let's be clear about the nomenclature of this business. Words such as *agent* and *broker,* often used interchangeably in popular speech, actually have distinct meanings.

- An *agent* is authorized to act on behalf of another person. A real estate salesperson is actually an agent of the broker, not the buyer or seller. While this may seem like a trivial distinction, it becomes critical when getting into issues of representation.
- A *broker* represents one or, under certain arrangements to be discussed in Chapter 6, both parties in a transaction. The broker has different licensing requirements than a salesperson. Even while you might be working directly with a salesperson, for legal purposes it is the broker who represents the buyer and seller in the transaction.
- REALTOR® is a registered trade name to be used only by members of the state and local real estate boards affiliated with the National Association of REALTORS® (NAR).
- A *salesperson* or *sales associate* is employed either directly or indirectly by a real estate broker to perform various services related to the transaction. A salesperson may be licensed as a broker (broker-associate) but still work under the designated broker, who is the broker-owner or broker-manager.

In this book we'll use the term real estate agent generically to mean the salespersons involved in the transaction, regardless of the fact that they may be licensed as brokers.

There are about 2.5 million real estate licensees in the United States. The best agents usually will be among the 700,000 members of the NAR. No agent can use the designation REALTOR® without being a member of the association, and hence of their state or local board. REALTORS® subscribe to a professional code of ethics and are also afforded many educational opportunities. Though some non-REALTOR® agents may be top-notch, in general the full-time REALTORS® tend to know the most about the local market. That's important because much of the service an agent offers relates to the quality of their information and knowledge. NAR, through its affiliates, awards designations that are earned through advanced coursework, experience, and elective credits. The Residential Sales Council offers the Certified Residential Specialist (CRS) designation; brokers can earn the Certified Residential Brokerage Manager (CRB) designation from the Real Estate Brokerage Managers Council.

An agent should be an expert on a market. Agents have access to the Multiple Listing Service (MLS), a database of properties listed for sale, and can obtain information about properties throughout a wide area. The agent also may be a good source of information about the community—schools, churches and synagogues, zoning regulations, development trends, municipal services, and more. In addition, agents also can help you with all the hurdles along the way. They may advise you on your choices in the mortgage market and recommend lenders. Since the agent's commission depends on a successful closing of the sale, the agent will help make sure that all the necessary paperwork is taken care of, including title insurance checks, home inspection reports, and mortgage commitments.

While agents have a clearly defined role in the real estate transaction, they also may serve as consultants in your home purchase. It's likely you'll spend lots of time with the agent. More than any other professional you work with, you'll get to know that person. You may drive around town together or sit in traffic jams. So choose somebody you can get along with. Just remember, your agent's primary mission is to help you find a home and to make the sale.

Attorney

Laws and customs governing real estate transactions vary from state to state. Some require that you use a lawyer, many do not. Regardless of the conditions in your state, hire a lawyer. This is good advice for any homebuyer but particularly a first-timer. Buyers describe a real estate closing as a flurry of documents flying back and forth over the table. With sweaty palms they sign here, initial there after a brief description (five words to one sentence) of what each document means, which they can't really concentrate on anyway. Won't it feel better to be accompanied by a knowledgeable expert who represents you in the transaction and has reviewed all the documents ahead of time? Don't try to save money by not using a lawyer. You have too much to lose. In the grand scheme of things, considering your level of risk, you won't be saving all that much. Protect yourself from surprises by hiring a lawyer who charges a flat fee rather than by the hour. Lawyer fees can range from a few hundred dollars up to $600 or more, varying according to the level of services provided and where you are. Still less than, say, hiring a painter for your new house. If you must save money, do the painting yourself; but don't be your own legal representative.

It's wise to involve a lawyer as early as possible, ideally before you sign a purchase agreement, so you can truly benefit from the legal counsel. Lawyers will advise you on the purchase agreement, and they also will prepare or review all the closing matters and documents, such as title policies, surveys, and closing statements. Don't try to save money by hiring a lawyer only for the close. There won't be time for the lawyer to examine the details, and problems might go unnoticed. And even if they are caught, do you really want to deal with an easement problem, for instance, at the closing table?

You should use a lawyer who specializes in real estate. Not that this area of law is especially complicated. However, there are a myriad of details. A lawyer who specializes in real estate will run a more efficient operation. Simple things like knowing whom to talk to in the recording office can make all the difference. You're likely to get better service at a reasonable cost from a real estate lawyer.

Shopping for a lawyer can save you some money. Ask friends and relatives for recommendations, as well as your real estate agent. Sometimes agents refer lots of clients to a lawyer who will offer an attractive rate in return. Lawyers' fees also can be negotiable. If not, try another lawyer. There are plenty eager for the work.

Home Inspectors

Most purchase offers are contingent upon a professional inspection of the home. If major problems are found, you have the option of backing out of the deal. Problem areas that will need attention, but are not serious enough to make you change your mind on the purchase, become negotiating points. You should definitely hire a professional to inspect the home before your purchase. It may keep you from making a big mistake or save you money. If no major problems are found, you will feel reassured. You also will have a written report that you can use as your maintenance plan for the first couple years of ownership. The inspection will probably cost around $250 to $350, but it's worth every penny.

Typically, the purchase agreement will define a time frame for the inspection, so you'll want to have conducted your search before you make an offer. Overall, home inspection is not a regulated industry with licensing requirements although some states require licensing and bonding. The best you can do is to ask inspectors if they belong to the major professional association, American Society of Home Inspection (ASHI), whose members subscribe to a code of ethics, are experienced, and have many continuing education opportunities.

You should ask friends for recommendations. Real estate agents also are good resources. Ideally, you should ask agents other than those involved in your transaction. If an agent is funneling a lot of referral business to an inspector, it's natural that a sense of loyalty will develop toward the agent. Even honest inspectors will have that in the back of their mind, and the less than ethical ones may deliberately fudge the report so that their friends' deal does not fall through. Let there be no doubt that the inspector is working for you, not the real estate agent.

☞ **Money-$aving Tip #15** *The ASHI has a direct response fax-back service that offers a list of member inspectors in your area. If you don't have a fax machine, ASHI will mail the report to you. The number is 800-743-2744.*

Steer clear of inspectors who are also contractors. They may be using inspections to generate new business. Along with your report, you'll get an estimate for repair and their business card. This is another obvious conflict of interest.

☞ **Money-$aving Tip #16** *If you spot an obvious and substantial repair expense, you may want to call in a specialist who can give you an estimate before you make an offer. For instance, a dead tree in the backyard may oftentimes be overlooked by a buyer but would nevertheless be a significant expense to remove. A landscape contractor can tell you exactly what it will cost.*

When you have your short list of recommended inspectors, interview them over the phone. Ask about their credentials. Sometimes retired contractors become home inspectors. Their wealth of experience is quite an asset. An inspector with an engineering background is also desirable, particularly one in structural engineering. Ask for references of customers they have served in the past six months. If they refuse to supply this information, you should be suspicious. Make calls to the references and ask how their inspections went, if any problems were found, or if any problems were missed. Confirm that the inspector will supply you with a written report. Let the inspector know that you'd like to be present at the inspection.

Mortgage Brokers

In Chapter 2, we looked at the loan application from your lender, which could be a bank, savings and loan, credit union, or one of several other financial institutions. You may choose to go directly to lenders or with a mortgage broker, who is an intermediary between you, the customer, and the lending source. The broker performs a marketing function for the bank. The mortgage banking industry can be competitive. Larger banks advertise heavily in local media. One way the lender can attract customers is to offer wholesale rates to brokers out there working with homebuyers like you. The broker makes money by retailing the loan to you. That doesn't mean it costs more to go through a broker. It shouldn't, not unless they are padding their fees, which is something you should look out for.

☞ **Money-$aving Tip #17** *Mortgage brokers can assist homebuyers with past credit problems, high qualifying ratios, or little money for down payments. Because they deal with a wide range of lenders, they'll know which affordability programs are available.*

Brokers may actually save you money. They will help you find the loan that suits your needs. Because some brokers deal in loans from 500 or more lenders, they tend to offer loans with competitive rates.

☞ **Money-$aving Tip #18** *Mortgage brokers may prequalify you, just as lenders do.*

Mortgage brokers have access to many types of loans, so they can be especially helpful if you have difficulty qualifying for a loan. If a mortgage broker succeeds in obtaining a loan for you, he receives a fee. But because they do not get a fee unless a match is made, they will work hard in your best interests. Be certain that the mortgage broker is reputable and that the mortgage divisions of commercial and savings banks will accept loans from that broker without assessing you additional fees.

Appraisers

Though the home you buy will likely get a full appraisal, you will not deal directly with the appraiser. Most lenders require an appraisal at your expense. Besides paying the fee or exercising your right to see the appraisal, you will have no contact at all with the appraiser.

However you may choose to hire an appraiser before you make your offer. An appraisal can cost roughly $200 to $400, depending on where you live. Of course, this step will slow you down a few days, and in a seller's market you risk losing out on the house. Hiring your own appraiser should not be necessary, though you may want to consider one under certain circumstances.

When you zero in on a house that you want to buy but are unsure of how much it's really worth, getting an appraisal can resolve your doubts and give you an accurate basis for the price you pay. But use a certified appraiser. Look for someone who is a designated member of a professional appraisal association, such as the Appraisal Institute, the American Society of Appraisers, or the National Association of Independent Fee Appraisers. Appraisers are listed in the yellow pages. A good idea, especially if you know where you'll be getting your mortgage, is to ask a lender for a recommendation. Ask also if you can use the appraisal report for your mortgage approval, and hence save several hundred dollars. Lenders have a vested interest in conducting their own appraisal, though, so they will be awfully careful about what appraisal reports they want to accept. Depending on your circumstances, your relationship with the lender, and where you are in the mortgage process, they may say yes. It's worth asking.

☞ **Money-$aving Tip #19** *If you're convinced that a house is overpriced but the owner is stubborn about coming down, you might want to use an appraisal as a negotiating tactic. Of course, there's always the chance that the owner still won't budge . . . or that the appraisal will prove the owner's pricing to be fair.*

Insurance Agent

You will need to buy hazard or homeowners insurance when you purchase a home. Later in the book we'll consider the particular factors to look for in insurance. What's most important is to get the right policy for your needs at a good price. If you already use an agent for automobile or life insurance, you should ask if the agent deals in homeowners policies and get some information. You might save money by using the same company.

Accountant

You certainly don't need to go out and hire an accountant because you're going to be buying a home. However, if you already use one, you probably should discuss your plans. The accountant will be able to give you specific information on how a purchase will affect your tax picture. If you're self-employed, your accountant will be particularly helpful in getting together documentation for lenders. Give the accountant as much advance notice as possible to be sure that your business records will be in good order.

How *Your Agent Can Help*

The real estate agent will be able to tap into a Rolodex file chock-full of professionals in these specialties. Don't hesitate to ask for recommendations. Sure, there is always the possibility of a conflict of interest, as noted with inspectors. In that case, you can ask for a few names or use other sources in addition to your agent. It's in the agent's best interest to refer you to people who will serve you well. If you're happy with the way your transaction was handled, you'll send other people to the agent.

The agent also can offer an alternative to an appraisal, the *competitive market analysis.* One of the methods appraisers use to determine value is to look at the sales prices of comparable homes. Real estate agents do the same analysis to help sellers set their price. They also can do a competitive market analysis for a home you're interested in bidding on.

Commonly Asked Questions

Q. What if I don't like my real estate agent? Can I change?

A. It depends. In Chapter 6, you'll learn about the different representational roles or agency relationships in real estate. If your agent is a buyer's agent you may have a legal commitment to stay with that person. You will sign an agreement outlining each party's obligations to the other, and your options should you want to change will be covered. In traditional real estate transactions agents represent the seller during the negotiations, even though they have been working with the buyer and brought the buyer to the property. In such a case, you have no legal commitment to stay with an agent.

However, it does make sense to stick with one agent. Over time the agent will become more aware of what you're looking for in a home and will be able to offer you better service. It's one thing if you're not getting along, but you're not really serving yourself by changing agents willy-nilly.

Q. Will a home inspector be able to identify environmental hazards?

A. This is a question to ask the inspector. To a degree, it depends on the property you're considering. Inspection for some environmental hazards, such as asbestos, requires special certification. On the other hand, asbestos insulation around pipes in an older building can be so obvious that any lay person would be able to identify it. If you want to get an accurate cost of removal or abatement of an environmental problem, you may have to hire a specialist. A good inspector probably can alert you to possible red flags in environmental issues. If you suspect serious environmental problems, it's probably smart to make sure you hire inspectors with sound credentials for identifying them.

Q. Isn't it a better deal to work directly with a bank than with the mortgage broker?

A. The interest rate and terms of any loan are identical whether you get the loan directly from the lender or from the broker. The mortgage broker performs a shopping function for you and is an expert on what choices are available in your local mortgage market. So working with the broker could turn out to be quite a good deal. You should ask a broker for a full breakdown of fees related to the mortgage, just as you would with a lender. Avoid mortgage brokers who require up-front retainers.

CHAPTER 5

All Types of Places to Call Home

The majority of homebuyers are looking for a single-family, detached home. But this is not all of America's dream. You'll have other choices. For many, a condominium (condo), town house, or cooperative (co-op) better suits their lifestyle or their budget. This is the choice of quite a few first-time buyers. Well-heeled singles might yearn for the tax benefits of home ownership, but still desire a lifestyle akin to apartment living. Why maintain a lawn when you don't have children to enjoy it? The thought of spending a Saturday puttering around the house fixing things may turn your stomach; perhaps playing tennis or lounging by a pool with neighbors is closer to what you have in mind. Condominium living holds attractions for all kinds of buyers. The single parent, for instance, might appreciate the stability of owning property in a solid school district but doesn't have the time for maintenance projects. On average, a condominium or town house also will be more affordable than a single-family, detached home.

While we associate the words condo, town house, or co-op with a specific type of housing, they actually refer to a form of ownership. When you buy a condominium or town house you also buy an interest in common elements of the facility or de-

velopment. A cooperative, though also a type of shared ownership, is quite distinct in its legal makeup. You'll see co-ops in
New York, Chicago, and a few other large cities. They are not
common anywhere else. Another choice is to share space but
not ownership; that is, be the landlord in a multifamily, owner-
occupied rental.

If you're set on a single-family, detached house, you still
have to choose between a resale home and new construction.
In this chapter, we'll look at new home construction, including manufactured housing.

While condos, co-ops, and town houses are rapidly cropping up around the country, the vast majority of buyers, about
85 percent, are still looking for a single-family, detached, existing home, which is the focus in this book. Most of the information, however, also applies to other types of housing. In this
chapter, we'll look at the special characteristics of those alternate choices.

Common Ownership Models

The option of purchasing a condo, town house, or cooperative can be quite appealing for the first-time buyer. For one,
the average cost is less than buying a single-family, detached
home. This is not to say that they are necessarily inexpensive.
A well-appointed condominium in a prestigious central city
neighborhood is sure to cost you big bucks.

☞ **Money-$aving Tip #20** *While condos, co-ops, and
town houses cost less and are therefore easier to get into, their
track record shows that they do not appreciate as fast as single-family, detached homes and will probably take longer to
sell.*

One of the biggest burdens of home ownership is maintenance. Common ownership lightens your load considerably.
You will only be responsible for maintaining the interior of the
unit. No lawns to cut in the summer, no driveways to shovel in
the winter. No climbing on ladders to caulk the second floor
windows. Maintenance-free living is an obvious attraction for

busy professionals, people with disabilities, young people, and senior citizens.

We have talked about how in owning a home you become more firmly a part of a community. As you can imagine, this is especially true under common ownership models. Many condo buildings or town house developments feature recreational and social amenities that foster community. There may be a hall available for social events, or the association may sponsor events for the whole community. Many developments also include swimming pools, tennis courts, or golf courses.

Nothing comes without its price. Unfortunately, it may be that you get to know your neighbors not poolside, but rather at association meetings, haggling over home association business, rules, or controversies. Homeowners associations have budgetary responsibilities. Each unit owner will pay a monthly fee, which is applied to cover maintenance, insurance, management, and other expenses. Fees are designed to pay for ongoing maintenance; however, sometimes special assessments must be made to cover the cost of a major repair. These are subject to votes by the board or members of the association. The homeowners association also will set rules. For instance, there may be restrictions on changes you can make to the outside of your town house, even on such details as the color of paint you use. There may be restrictions on children and pets. Wherever money is involved, not to mention rules affecting lifestyle, you can expect plenty of opportunity for disagreement and controversy.

☞ **Money-$aving Tip #21** *Ask the opinion of your lawyer or an insurance agent as to whether the insurance coverage of the homeowners association is adequate.*

Condominiums

In the condominium form of ownership, the owner of the unit also owns an undivided interest in all common areas, such as hallways, parking lots, and land. The share is called "undivided" in that it is not a clearly defined portion nor does it bear any relationship to the size or value of the individual unit. The owner's share in common areas cannot be sold separately from the unit itself. Ownership of the unit includes everything within the outside walls and floor and ceiling. In effect, the owner has rights to a horizontal layer of airspace.

Each unit of the condo development is a separate entity that can be mortgaged, taxed, and sold separately from all others. A unit also can be foreclosed separately from all others in case of default. State laws require the developer of the condo project to execute a master deed accompanied by a *declaration.* The declaration will address the legal rights and obligations of owners, limitations of use of general areas or even units (such as number of occupants), the bylaws of the association, procedures for determining and collecting fees, voting rights (oftentimes weighted according to the size or value of the unit), and more. It often runs longer than 30 pages. Bylaws may be changed by vote of the entire membership. Consult your real estate agent, or better yet your attorney, if you have questions. Obviously, this is an important document that you should understand completely before committing to buying into a development. You never know what you'll find in the small print. In some condo developments, the resale of your unit may be subject to the right of first refusal of other owners.

The homeowners association will have a budget that covers the fees assessed to the unit owners for maintenance, repair, and improvements, and for the expenditures that have been made. As a purchaser you have the right to look into these records. You should check that all the owners are current in homeowners association fees, and review the procedure if these fees are not paid. Though the homeowners association does not directly suffer if a unit owner defaults on a mortgage, effectively it does. More than likely the defaulting buyer will not be current on homeowners association fees. The association can put a lien on the deed, but it will be secondary to that of the lender. Most likely the association will never see the money, and the fees will have to be made up by the rest of the owners.

☞ **Money-$aving Tip #22** *Watch out for homeowners association fees in new developments that seem particularly low. Sometimes the developer of a new project will use low fees to attract buyers; however, you may be due for a hefty increase in the near future.*

The homeowners association is a legal entity. As a member, you share in its liability. If the association is named in a lawsuit, you are collectively responsible for any judgments or settlement costs. Many homeowners associations carry insurance policies to protect against this and other similar risks. You should check into the extent of the insurance coverage the homeowners association carries. Unfortunately in our society, it doesn't take much, sometimes not even just cause, to trigger a lawsuit.

Location, of course, is always important in real estate, and particularly so with condominiums, which are most popular in urban and resort areas, where usable land is relatively scarce. Condo buyers are usually looking for amenities. Even if you don't feel a particular need yourself, you must be sensitive to this desire so you're not stuck with a hard-to-sell condo unit. Here are a few things to look for:

- Good or adequate parking, including some accommodation for guests of unit owners
- Recreational facilities, such as parks, swimming pools, tennis courts, golf courses, or health clubs, either on-site or nearby
- Convenient to centers of employment or transportation (though not adjacent to a major road)
- Nearby or within a neighborhood of predominantly single-family homes

One drawback of condos, particularly in highrise developments, is that they can be noisy. You share walls with your neighbors and the level of soundproofing within these walls can vary dramatically. Generally, units built as condos, rather than converted apartments, will be quieter. Regardless, be sure to listen too while you're looking at condos.

You also should be aware that some condo projects are primarily occupied by renters. As condos were built many investors bought up multiple units to use as rentals in anticipation of selling a few years down the road for a profit. It's typical that a few owners also, because of a job relocation or whatever, will end up renting their units. However, when renters dominate a condo project, occupying owners could suffer. Renters are less concerned with upkeep of the property. The poor condition of other units in the building will hold down the value of your own. Many lending institutions have rules that prohibit them from making loans on condo developments where the renter to owner ratio is high.

Town Houses

Town house developments generally follow the condominium form of ownership with an important variation. Each owner controls part of the actual structure of the buildings— the floor, the roof, and a share of common walls. Unlike condos, however, the owner also enjoys air rights; that is, ownership of the ground below the unit and the air above it. In a condominium you are likely to have neighbors above and/or below your unit. The town house is closer to a single-family home in its structure. Most are two-stories tall, with bedrooms on the second floor and living areas on the first. Descendants of the row houses popular in the urban centers of America and Europe a century ago, town houses will share common walls. The surrounding land of the development, recreational facilities, and all common areas are owned jointly by the owners of all units.

Town house developments by their nature are less dense than condo buildings. You may have the benefit of a patio, garden, or small yard, but you won't have the responsibility of cutting the grass, which would be handled by the homeowners association.

Cooperatives

Though not all that common nationwide, cooperatives are an option in large cities like New York or Chicago. In terms of housing style, cooperatives are similar to condominium developments. However, the form of ownership is quite distinct. When you purchase into a cooperative you get stock in a corporation and a proprietary lease granting tenancy in a building. The corporation owns the building and may have a mortgage on the property. The members of the cooperative pay a pro-rated share of the corporation's expenses, such as mortgage payments, real estate taxes, maintenance, and payroll. A shareholder may freely assign the stock certificate. Assignability of the proprietary lease, however, is severely restricted. You may not be be able to rent out your cooperative unit. On the plus side, neighbors in a cooperative will most likely be owners, not tenants (ask for the number of owner occupants versus the number of tenants).

Cooperatives often require substantial down payments and financing can be more difficult to obtain than for a condominium. Shareholders do not actually own their units, and therefore are subject to more restrictions than condominium owners about how the units are used. Because the cooperative owns the building, if an individual shareholder defaults on a mortgage or taxes, the corporation must make up the difference or risk losing the entire project to foreclosure or a tax sale. To protect themselves against this risk, many cooperatives require a prepayment in order to build up a reserve fund.

As you might guess, this form of ownership creates an environment where shareholders can be picky about who buys into the development. You will need to be approved as a potential purchaser by the cooperative's board of directors. And when it comes time to sell, your purchaser will undergo a similar review. You can pretty much forget about selling your cooperative unit to a drummer in a rock and roll band.

☞ **Money-$aving Tip #23** *If you anticipate a short term of ownership, a cooperative is probably not a good option.*

Owner-Occupied Rental Properties

Another option in home ownership is to be homeowner and landlord in a duplex or two-unit residential property. If you have more money to invest (and are ready to be more of a landlord), you might consider larger buildings of three or four units or more.

For the most part, your financing options for owner-occupied rental properties are similar to single-family homes. You will pay a slightly higher rate and FHA and VA loans are available. It is key in the lender's eyes that you occupy the building because its risk is substantially less with the owner on the property. In addition, many jurisdictions will tax income properties at a lower rate if the owner occupies the premises.

Operating a residential, income-producing property, even one as small as a duplex, is a business. You should therefore look at properties with a business perspective. For starters, you should be certain that the area is zoned for multifamily residences. This shouldn't be a problem in areas where the homes were originally built for that purpose; however, in some areas large houses have been remodeled and converted to multiple units.

The seller of the building will provide you with the current rental income. Study ads to see if rents are up to the market rate. Long-time tenants may very well be paying less. Assuming property values have been increasing in the area, you'll have a larger mortgage to pay than the current owner and may therefore need to raise the rent. You'll risk losing tenants if your rent increase is substantial, so you have to factor in the costs of securing new tenants and the possibility that a unit will be vacant for a couple months. Study the larger market from a renter's perspective. Does the neighborhood offer the amenities that renters are looking for? For instance, renters that are more likely to be single may appreciate proximity to restaurants, bars, and clubs. Amenities such as nearby shopping or public transportation are also important. Just like homes and most everything else, rents are subject to supply and demand. If large new apartment buildings are under construction in the neighborhood, they'll have an adverse effect on your income potential. If apartments in the area are renting fast, you'll be in good position.

Lenders will take into account your rental income when determining your qualifying ratios. However, they will probably not credit you with the full amount. To account for the potential of vacancies, they'll discount the rent and count only a portion, probably around 75 percent. They also may take into account the tax benefits, which can be substantial, when reviewing your application.

Owners of rental properties file a Schedule E with their income taxes, which itemizes income and expenses for the rental property. You are able to write off all maintenance and repair expenses that are specific to the rental unit, as well as any business expenses related to its operation, such as advertising. You also can deduct a prorated share of expenses affecting the building. For instance, if you hire painters for the exterior of a duplex, half of the cost is deductible. Depreciation too is an important benefit available only to income-producing property. Depreciation takes into account the physical deterioration of the building and its systems, which over time lessens the value of the property. The owner of a duplex can depreciate half of the building's value according to the IRS's "straight-line" method. (Land does not wear out, and therefore cannot be depreciated.) For details, check with your accountant. What's great about depreciation is that it's a "paper loss," involving no out-of-pocket expense. Most owners of duplexes will operate at a loss, particularly when depreciation is factored in. The investment is therefore a tax shelter, offsetting its income as well as some ordinary income. Good recordkeeping is critical for a duplex owner. When it's time to sell, there are essentially two properties being sold (primary residence and rental) with very different tax treatments.

So the small landlord enjoys significant benefits, with income to help pay the mortgage and extra breaks at tax time. Keep in mind, however, that it is a job and will take up your time. A tenant is your customer. When there is a maintenance problem, you will need to take care of it immediately, whereas in your own home you might be able to put it off until you have more free time.

☞ **Money-$aving Tip #24** *Those who are handy and can take care of repairs and remodeling without hiring expensive tradespeople are best suited to being landlords.*

New Houses

You'll hear strong opinions on both sides of the new home versus resale debate. There are those who wouldn't be caught dead in a new home development. They prefer to live in established communities with a tradition of housing, schools, churches and synagogues, and shopping that has evolved over time, rather than at the drafting table of architects and planners. They seek a charm that may be reminiscent of their child-hood neighborhood. Even if driven by emotions, they conjure a lost idea of craftsmanship with pithy phrases such as "They don't build them like they used to." For others, new homes of-fer the best and latest of homebuilding technology. They want to handpick the amenities they find most important, appreciat-ing the degree of customization available even in new home developments. Many young couples look forward to moving into a community with others like themselves who are just starting their families. And the quality idea they hold to is "It's brand new, we won't have to fix anything." In the end, the right decision is largely a matter of taste.

Builders

By far the most important factor in selecting a new home is to choose one from a reputable builder. Many new homes come with warranties that insure the property against major structural defects. Buyers shouldn't count on such warranties to make up for a shoddy or dishonest builder. The home war-ranties are designed to come through when the builder can't or won't pay. Most new home complaints are resolved directly with the builder. So the builder's willingness to make repairs is more important than any warranty the home carries. It is not unusual for builders and warranty companies to deflect liabil-ity toward one another, leaving the homebuyer in the lurch while they sue each other. Some builders skip town alto-gether. The builder who doesn't take care of the customer is not much better.

So it pays to check out a builder's reputation and track record. It's old advice for sure, but nonetheless wise—you should talk to others who have bought from the builder. Call the local better business bureau to see what, if any, complaints were lodged against the builder and how they were addressed. Look for builders who have been in business for a long time un-der the same company name and steer clear of any firms that are incorporating under different names and entities. You also can check if a builder is a member of the National Association of Home Builders. Membership does not ensure quality, but it is at least some assurance of a builder's standing in the profession.

☞ **Money-$aving Tip #25** *Most, though not all, reputable builders use their family name for the company. Scrutinize carefully those companies who promote a development under names like River Cascades with little or no mention of their own name.*

Quality builders take pride in their homes. They do things right the first time because it costs more to go back and make repairs. When there are problems, they give them prompt attention. They value satisfied customers. You'll do more to protect yourself against defects by choosing your builder wisely than by supervising construction (unless you happen to be a structural engineer or professional contractor yourself).

Buying a New House

When you shop for new homes, you most likely will be visiting show homes. And they are showy. Remember to keep your eye on the substance. Many show homes are exquisitely decorated to spark your imagination, but you should look through that and focus on design and quality. When you choose your model you will be selecting various features. You need to understand what in the show homes is included in the basic package and how much you will have to pay for the enhancements you desire. No sense in being allured by the master bedroom's bathroom if you can't possibly afford its amenities in the model you choose. Salespeople have a label for this tactic—the bait and switch.

Take a look at the plans for the various models you are considering and ask for explanations if anything is unclear. If at all possible, try to see your actual model. You also should ask the make and models of all appliances and the manufacturers of various components, such as windows and plumbing fixtures. Demand quality. It will be worthwhile to pay a little extra for top-line fixtures and appliances. In the long run, they will pay for themselves. As always, to avoid misunderstandings or outright deceit, it's best to put in writing what exactly you'll be getting.

New homes tend to be a little more expensive than resales. However, unlike the 1980s when luxury homes were the fashion, builders are now paying more attention to affordable housing. Many new developments are cropping up in the exurbs, on the ever-expanding limits of the metropolitan areas. First-time buyers are finding they can get more house for their money, or perhaps just a starter home they can afford, if they're willing to move farther out from the city center.

You'll have less negotiating room with a builder. In the early 1990s the National Association of REALTORS® tracked actual selling prices and found that resale sellers accepted a median drop of $4,000 from the asking price, while builders trimmed only $500. Even if units are selling poorly in a project, builders may be reluctant to lower prices for fear of alienating customers who have already purchased. Builders also may refer you to lenders or mortgage brokers who can offer you good rates, sometimes even below market. Another way builders will help buyers get in is by covering some of the closing costs or paying the points on the loan.

☞ **Money-$aving Tip #26** *When negotiating with builders, look for financing incentives rather than price reductions.*

Just because a house is new is no assurance that you will not have maintenance problems. In fact, in a resale you can expect that any major structural problems have already been addressed. You may have heard horror stories about major problems in new construction or reports about less serious defects that prove to be a constant nuisance. Builders also are notorious for construction delays, which can wreak havoc in coordinating your move with your landlord. It's not uncommon for buyers to have to take refuge for a month or two with friends or relatives or at a local motel while awaiting completion of their home. You should therefore persistently track progress on your new home construction and plead for flexibility from your landlord.

Don't overlook the inspection process either. You may want to hire your own inspector for the final walk-through. If you still have last minute work to do, make sure it's completed before the closing of the sale. The builder will not be as motivated to complete the work once the deal is done. You might let small touch-up work wait until after the close. But whatever you let slide, be prepared to do the work yourself because you might have to.

While not without challenges, new homes offer buyers the opportunity to take advantage of the latest advances in home construction technology. While it's true that the attention to craftsmanship is not what it used to be, new materials can be superior. Drywall, for instance, will not crack the way plaster does, and when repairs are necessary they are inexpensive. You'll also get homes that are designed for the way we live today. The homes of the past, for example, were not wired for people with home offices, microwaves, and elaborate television and stereo setups.

☞ **Money-$aving Tip #27** *Watch out for hefty property tax increases when buying a new home. The current property tax rate is probably based on the value of the vacant lot. The exact amount you will be paying when your home is part of the assessment may or may not be established at the time of the sale. Ask your builder or agent about this. Your taxes could easily go from a couple hundred to a couple thousand dollars.*

Build-It-Yourself and Manufactured Homes

It's possible that you could buy a lot and build your own home. Unless you're a professional contractor or have significant expertise in the building trades, it's probably not a good option for the first-time buyer. For one, financing the project can be tricky. You should pay cash for the lot itself. Even if you can swing that, you'll have other obstacles. A bank or other mortgage lender cannot legally issue a mortgage loan on an unfinished house, so financing is a two-step process. First you will need a loan to pay for the building materials and construction. The second step is to convert the construction loan to a standard mortgage loan when the house is completed. When you go into banks as an amateur or one-time homebuilder, officers probably won't even want to talk with you. The risk is too high. They imagine being stuck with an unfinished house or messed-up construction. If you really want to take a stab at this, you'll need to hire an architect and a professional contractor.

A more realistic and economic option along these lines is to buy a *manufactured house,* which might go by various other names—such as *factory-built, prefabricated, prefab,* or *mobile home* (a term the industry dislikes since it refers only to the most modest and basic form of manufactured housing).

The manufactured house has a long history. Sears Roebuck sold them in the 1930s. Even earlier, the distinguished Frank Lloyd Wright designed factory-made homes for the working-class Americans who could not afford the conventional, "stick-built" construction. Manufactured homes are built to varying degrees in factories and then transported to the construction site. Modular homes are 95 percent built when they leave the factory; at the construction site little more is done beyond hooking up utilities and mechanical systems. Panelized or pre-cut homes require more work at the site.

Today, manufactured houses can be quite luxurious. You can get virtually any kind of manufactured house, from the one-bedroom mobile home for $25,000 to models going for as much as half a million dollars. More likely, your price will fall into the $50,000 to $250,000 range. Quite a few builders in new home developments use manufactured housing. About 40 percent of new homes built today are actually manufactured houses. The controls and standardization of assembly-line construction can make manufactured houses a better bet for quality than their stick-built counterparts. Materials are often top-notch, and so is the design. Don't forget, because houses are transported, possibly even lifted up and dropped on a site, they are made to be durable. Design is also of high quality. According to the American Institute of Architects, only 15 percent of conventionally built new homes are designed by professional architects. It is more common for manufacturers to hire certified architects.

If you're interested in manufactured housing, an excellent resource is *Manufactured Houses: Finding and Buying Your Dream Home for Less,* by A.M. Watkins (Dearborn Trade, 1994). The book includes a directory of home manufacturers.

How Your Agent Can Help

Unlike buying a single-family, detached home, when you purchase a town house, condominium, or cooperative, you're also entering into a legal and business relationship with your neighbors. It's more than just buying a property. You're joining an association, and one that may not be so easy to get out of (depending on how fast units sell). You should ask your agent about the issues raised in this chapter.

Agents may know quite a bit about a condominium project. Ask if they've sold any units. If they have, they can tell you how long units are generally on the market and approximately what the average selling price is. Ask if the agent has seen the declaration and bylaws. Have there been disputes in the past between association members? Ask about the culture and lifestyle of the community. Remember, agents may be reluctant to discuss social characteristics because of racial steering laws.

If you're interested in new construction, don't feel that you have to work directly with builders or their sales agents. Not all builders have their own sales associates and many who do also cooperate with area agents. In many cases, your agent can represent you just as in the purchase of an existing home. Ask agents about trends in new home developments in your area. In new home construction, the builder is as important as the construction itself. Your agent may be able to recommend reputable companies.

Commonly Asked Questions

Q. Do new home developments have homeowners associations?

A. Homeowners associations are legal entities organized around the declaration and bylaws. New homes that are part of a subdivision that shares property, such as a planned unit development (PUD), would have a homeowners association. Quite frequently, a new home development will have a homeowners association. If not, they almost always have conditions, covenants, and restrictions. But homeowners associations should not be confused with neighborhood groups, block clubs, or similar community organizations, which tend to be social or political in nature and do not in any way address legal rights of ownership or create binding regulations.

Q. If my homeowners association carries insurance, do I still need insurance for my unit?

A. Yes. The homeowners association insurance will only cover the common areas of the project, not individual units, which are owned by members and therefore are the responsibility of members.

Q. Would I be dealing with the builder or a real estate agent when buying a new home?

A. Builders use real estate agents to list a property or employ their own agents. You also might be dealing with a sales associate in the employ of the builder, or perhaps even directly with the builder. The agent you've been working with can contact a builder and arrange to show you model homes. Builders may or may not be willing to pay agents a commission. Some builders have sign-in sheets at their show homes. If you sign in, you may be acknowledging that the builder brought you in, essentially releasing the builder from paying your agent a commission.

CHAPTER 6

Working with Real Estate Agents

We all carry preconceptions about salespeople. The cynic says, "Never trust anyone on commission." Some of us will avoid even the fresh-faced, smiling sales clerk kindly offering us assistance at The Gap. The manipulative tricks of automobile salespeople are so notorious that a successful new company has sprung up built on, among other things, the promise not to engage in such tactics. On the other hand, there are those who love dealing with salespeople, embracing the give and take, the psychology of the sale. Not surprisingly, salespeople themselves love being sold to, perhaps out of appreciation of technique. In addition, they recognize the salesperson is a great source of information and not a threat at all as long as you hold the wallet.

In real estate sales, like any other field, you'll find hardworking, knowledgeable, and honest professionals, as well as some bad apples. By now you know that a good agent can be a huge help in your transaction. To truly understand how an agent can help you, we first have to take a look at an area of real estate law called agency relationships. Don't worry, if any legalese slips in it will be accompanied by a plain English explanation. Understanding agency relationships is critical to the transaction so don't think about skipping over this part. You'll learn a bit about how agents go about their jobs. The goal is for you to make a reasonable judgment on what kind of help you need from an agent. Ideally, you should be able to tell whether a prospective agent will serve you well. You'll definitely know right off when you're not getting the service you deserve.

The Traditional Real Estate Transaction

Until recently, nearly all real estate agents worked for the seller. The agents showing the buyers houses and eventually presenting their offers to sellers were, in fact, representing the seller. In such a role, the agent would be called the *subagent* of the seller. The terms *cooperating broker* or *selling agent* also are used. Such an arrangement continues, but it is only one of your options.

Here's how the system has traditionally worked. The home-seller signs an agreement with an agent to sell the home. This agent is called the *listing agent,* who publicizes the property to other agents in a wide area through a database called the Multiple Listing Service (MLS). Other agents review the information in the listings and alert the buyers to homes that meet their criteria. Some will bring buyers to see the property. If the buyers decide to make a bid on the property, their agent, as a subagent of the seller, is duty-bound to obtain for the sellers the highest possible price. This arrangement may surprise you. Research has shown that a large majority of buyers, even after going through the transaction, believed "their" agent had been working for them, not the seller. Consumer advocates raised the red flag, and many states passed laws requiring written disclosure of who the real estate agents are representing in the transaction. Nearly all states now require such disclosure.

To see it all spelled out in black and white was an eye-opener to many buyers. The clamor from consumer pressure groups as well as some from forces within the industry have brought change. Buyers' agents have actually been around for some time and were common in commercial transactions. Recently, changes in the rules of the MLS system have made the residential side more accommodating to buyer representation.

With that background, let's look at some of the key concepts in agency relationships.

Agency Relationships

Representation begins with the concept of fiduciary, a relationship that implies a position of trust or confidence in which one person holds or manages the money or property of another. In a real estate transaction the broker or salesperson is a fiduciary to the principal, who traditionally was the seller but in the case of buyer representation is the buyer. The fiduciary owes the principal complete allegiance: loyalty, obedience, and confidentiality regarding sensitive information concerning the principal; full disclosure of relevant information concerning other parties; the use of skill, care, and diligence; and a full accounting for all monies. The principal usually will have just cause for a lawsuit if any of these duties are breached.

To see fiduciary duty in action and what could happen if you don't understand where it lies, consider this example of a traditional real estate transaction. Joe and Marie have been working with Nancy, a subagent of the seller. They've settled on a house for which the seller is asking $140,000. "We've decided to offer $120,000," Joe tells Nancy.

"Ahhh, Joe, that's a little bit low," Nancy says. "I think the Joneses are going to find that insulting. They might not even counter."

"Well I'm willing to go up to $125,000; that's the most we can afford," Joe says.

"And we want to leave room for negotiation," Marie adds, "so we'll start with $120,000." Nancy writes up a purchase agreement with Joe and Marie and then sets up an appointment with the Joneses to present it.

"He said he'd go up to $125,000 and that's all they can afford," Nancy tells the Joneses. "But from what they've said about their income, they could qualify for your asking price. We've been looking at lots of places priced at 150. So it's what they'll pay, not what they can afford, that's the issue." Nancy has honored her duty, fully disclosing what she knew about Joe and Marie's position and putting the Joneses in the best possible position to negotiate. She also did not tell Joe and Marie that the Joneses had already signed an agreement to buy their next house and needed to sell their home in the next few weeks in order to fund the purchase. Sure doesn't seem fair; and it's not if Joe and Marie think Nancy represents them.

When the broker/salesperson's fiduciary duty is to the buyer, the relationship is known as *buyer agency* or *buyer brokerage.* But buyer representation comes in a couple of different flavors. In an *exclusive buyer agency,* the brokerage companies do not take any listings and work with buyers only. However, many brokerage companies practice what is called *disclosed dual agency* or *consensual dual agency.* They can then offer buyer representation and still take listings. Suppose you sign on with a buyer's agent. After looking around you fall in love with a home that another agent in the same brokerage company has listed. The brokerage company through its agents has pledged a loyalty to both sides of the negotiation. The company may carry through the transaction provided it informs both you and the seller, in writing, of its *dual agency* status. In this capacity the two agents' fiduciary duty is slightly different with confidentiality running both ways. While each is loyal to the party represented, the agents can't disclose what the buyer might pay or what the seller might accept. Nor can they share any other confidential information that might affect the bargaining position of either party. Note that disclosure of dual agency is a critical requirement; undisclosed dual agency is fraudulent.

Let's look at how Nancy would work with Joe, Marie, and the Joneses under buyer agency. Joe and Marie tell Nancy that they will offer $120,000 for the Jones's home, listed at $140,000.

"That offer might be low," Nancy says, "but let's go in with it. The Joneses signed an agreement last week with a contingency on selling their home. They may take 120."

After Joe tells Nancy that he can go to $125,000 as a maximum, she says, "It's not for me to tell you what you can afford, but remember how we figured out the mortgage you could get. I'd bet you'd be able to get in the home at 140 if you'd want to."

"But then we'd be putting all our savings in," Marie says. "We want to keep a nest egg."

"If $125,000 is truly your ceiling, maybe you want to go in lower to give yourself a little wiggle room," Nancy advises. Joe and Marie decide to offer $118,000. When presenting the offer to the Joneses, Nancy says nothing about Joe and Marie's willingness or ability to go higher.

Now let's pretend Nancy is acting as a dual agent. Essentially she has absolved herself of any role in the negotiations beyond moving them along by communicating the offers or counteroffers. She may encourage a counteroffer from either party, because she is hopeful that they will come to terms. This is a fine line, though; she can't become an advocate for either party's negotiating position. She advises neither buyer nor seller how to decide on the terms of the purchase. She does not tell Joe and Marie that the Joneses may be particularly motivated. She does not tell the Joneses what kind of counteroffer Joe and Marie might accept.

Dual agency is a relatively new practice and still makes many in the business uncomfortable. Extensive training and monitoring of agents is necessary and liability exposure is broadened. In order to avoid this situation and still take listings, some brokerage companies practice *single agency.* If you decided to bid on a property listed by the company, the agency agreement with either you or the seller will be terminated and the negotiating business will be referred to another broker. Terminating contracts has its own liability risk and single agency is probably feasible only in very small companies. The single agency brokerage company may choose to show you company listings first, before signing a buyer agency agreement. Besides the problem of restricting the properties you view, you still won't have the representation you seek. Another possibility is that the single agency broker may choose not to show you company listings. This, of course, also restricts your choices. The larger the company's share of an area's listings, the more problematic single agency becomes. If you choose to use buyer representation, you should be clear on whether the brokerage practices dual agency, and, if so, how it is handled.

Still another role, the concept of *facilitator,* is developing in the industry. Many believe that fiduciary duty itself confuses the role of the agent and causes a lot of misunderstanding. After all, consider the traditional real estate transaction. The forces of human nature tend to bend the strict requirements of agency law. The agent you've been working with may legally represent the seller. However, having spent hours with you and shared your hopes and dreams, can even the most conscientious agent ignore that emotional connection and take the side of a seller who is quite likely a complete stranger? The facilitator is an intermediary in the real estate transaction, helping buyers and seller reach agreement but not advocating for either side. Proponents of the facilitator concept argue that it best describes what agents have done all along.

Finally, you should know that agency relationships are governed by state law. Custom and practices, therefore, vary around the country. Buyer representation has quickly grown in popularity. In some areas, more than 90 percent of transactions involve a buyer's agent. However, there still may be pockets of the country where buyer representation is not accepted, and some listing agents may be reluctant to deal with a buyer's agent, thus restricting your choice of property.

☞ **Money-$aving Tip #28** *Ask agents or others about common representation practices in your area.*

Buyer Agency Agreements and Compensation

In the traditional real estate transaction the seller signs a listing agreement promising a commission to be paid to the broker. If a subagent from another brokerage is involved in the transaction, the commission would be split. As the concept of buyer agency developed, the question of who pays the commission was at the heart of the controversy. Would it be fair that the seller pay a commission to the advocate for the buyer? "After being chiseled down, now I have to turn around and pay this guy!" Could buyers' agents honor their fiduciary duty when their compensation would be coming from the seller? In fact, payment of commission does not determine where agency lies. While it's true that sellers pay commission, the money is accounted for at the close, and it's buyers who are bringing it to the table. To make this truth more explicit, the industry often phrases the arrangement in a new way—the agent will earn a commission "from the proceeds of the transaction."

Just as a seller signs a listing agreement, if you decide to go with a buyer's agent, you will sign an agreement with the agent. The agreement defines the compensation, along with all the ground rules of the agency relationship. The agreement will probably require that you exclusively use that agent, who would be entitled to a commission even if you found the house you wanted without the assistance of the buyer's agent. It's possible that you could negotiate an agreement that is nonexclusive. Whether or not that's even desirable is subject to question. Presumably you've chosen buyer agency so you can have more help in the transaction. It stands to reason that agents will be more willing to put time into your house hunt if they can be assured that eventually it will earn them a commission. A potential problem with exclusive relationships is if you change your mind about areas and want to start looking in places where the agent is not active. The agent will still be able to negotiate the deal for you but may not be such a good source of information about the neighborhood and market trends. You might overcome this limitation by choosing a company with several offices in the area.

One key negotiating point is length of term. Of course, the agent will prefer a longer term, which increases the chance of a commission being earned. The buyer prefers a shorter term. If you're unhappy with the agent, you can move on that much sooner, no questions asked.

There are all kinds of arrangements for compensation. The good news is that you should be able to enlist the help of a buyer's agent without it costing you extra. Sometimes the agreement will specify that the agent will split the commission with the listing agent just as in the traditional arrangement. As buyer agency becomes more common there is less seller resistance to this. The buyer agent may request a small up-front fee or retainer, which is a deposit going toward a commission that the seller may pay. It works basically like a good-faith deposit. You may have to forfeit this money if you don't buy a house during the term of the agreement. Even if an agreement requires you to pay a 3 percent commission to the buyer agent, this should give sellers compelling reason to lower the price by 3 percent because they won't be paying a full commission to the listing broker. Customary real estate practices do tend to be local, so the best way to explore your options is to ask around.

Is Buyer Agency Best?

A strong case can be made that the best form of representation for the first-time buyer is buyer agency. Real estate agents know that first-time buyers will demand more of their time. When you sign an agreement to work with an agent, loyalty is a two-way street. The agent will feel more secure in spending the valuable time with you, just as you have the right to expect a higher level of commitment from the agent. Remember, all agents have to sell is their time and service. The best agents are able to prioritize their time, spending it where there is most likely to be a commission in the near future.

That said, you might still consider the traditional subagent relationship if it's practiced in your state. Just because you expect a higher level of service from a buyer's agent doesn't mean you'll get it. A good subagent beats a bad buyer's agent any day. You shouldn't misconstrue the discussion of agency relationships to mean that subagents are villains. The traditional system has worked well in the past, so long as buyers understood who the agent was working for. The seller's agent, though not a fiduciary, still has the following duties to the buyer:

- To be honest and deal in good faith
- To present all offers
- To disclose any known facts about the property that materially affect its value or desirability. Many states now have seller disclosure laws, and these requirements apply also to agents of the seller.

You should take a look at your circumstances when you make a decision about representation. Working in the traditional subagent system gives you more flexibility. If you're looking at houses in vastly distant parts of the city, you could even work with a couple of different agents. If you are dissatisfied with your agent, you can switch at any time.

Many Americans feel uncomfortable negotiating. It's not a big part of our culture. Some tremble at even the thought of it. If this describes you, then you should seriously consider a buyer's agent. However, plenty of buyers have done quite well at the negotiating table without an agent as their representative. You might even be a better negotiator than a lot of agents. Note that the agent should still be your go-between, presenting all offers. Emotions run high in real estate transactions. A poor choice of words when speaking to the seller could kill the negotiations. Even while maintaining fiduciary obligations, the selling agent can and will supply you with enough information about market conditions to put you in a position to bargain.

Homebuyers who are relocating to a different part of the country are prime candidates for buyer representation. You won't be able to dig around the market yourself. You will be unfamiliar with local real estate practices and will probably need extra assistance in lining up the various players who will carry you through the transaction. It makes sense to use a knowledgeable agent who is also your advocate.

In the end, the agency relationship is just one of several considerations. What matters more is that you work with an honest, knowledgeable professional.

☞ **Money-$aving Tip #29** *Look for an agent who is a member of the National Association of REALTORS®, as indicated by use of the trademarked REALTOR® designation. In addition to showing a commitment to the profession, members of the association also subscribe to a code of ethics.*

How Agents Work

Real estate agents enjoy a fair amount of independence. When looking for an agent, you should be sensitive to their business strategies—how they choose to work—as a way of evaluating how much service they will be prepared to give you. For instance, some agents want to concentrate on getting listings. They'd rather be presenting their services to potential homesellers than trekking around town with a buyer, especially a first-timer. Historically, the real estate business has attracted a lot of part-timers, though this is rapidly changing. A part-timer will have to make a convincing case that you will still get the same level of service. Somebody who works full time is demonstrating a stronger commitment both to the profession and to the customers. Finally, real estate agents establish all kinds of specializations. An obvious one is geographic area. But you could probably dig up a specialist around any other factor you can think of—price, type of property, expired listings, even first-time buyers. We're not suggesting that you need to find somebody working with only people like you, interested in similar homes and in the same neighborhoods. But you should be aware of what your agent's specialties are. If you're too far afield, you won't get the same level of service.

Real estate is not a nine-to-five job. Most successful full-timers put in more than a 40-hour week. Much of the activity takes place evenings and on weekends. Though you would like seven-day-a-week, 24-hour service, it's not humanly possible. So ask what systems are in place to give you service when you need it. Many of the top agents have their own personal assistants, who handle the many details of the transactions to free up the agent for selling and listing. Don't let this turn you off, as if you'll be slighted by being turned over to a lackey. On the contrary, this indicates a higher level of service. You'll have one more person to call. Some agents work in pairs as partners. This may allow the agents to work part time and still offer full-time service. Or it may be less formal, one agent allowing another to take a day or two off a week and still keep the business moving.

Finding Agents

The best way to find a good agent is through recommendations from a satisfied customer or client. If you happen to know several people who have bought in the neighborhood you're interested in, this might be relatively easy. If you're moving to a part of town where you don't know anyone, it's not that simple. Use your network. Let it be known to all your friends and relatives that you're starting a house hunt and where you plan to look.

Another good way to find an agent is to attend open houses. Agents conducting an open house will be more than willing to talk with you. Attending open houses is a great way to begin your search. You will get a feel for the market and at the same time can shop agents as well.

You can always just walk into a real estate office. Be sure you interview the agent thoroughly to ensure they'll fit your needs and look out for your interests. Of course, another option is to open the yellow pages, call around, talk to some agents, and see who you like.

How Well Are You Qualified?

In choosing an expert to work with, it is customary to have a list of questions with which you can screen prospective choices. What might be more revealing with real estate agents are the questions they ask you. Any good salesperson wants to qualify a lead. The scruffy teenager walking into the car dealership alone will get a different level of service than the one driving up in a luxury car with Dad. That's instant qualification. Agents should take more time, starting with basic questions to determine your motivation and ability to buy. The more successful they are in learning about you beyond these basics— your goals and desires, how you live your life—the better service they will be able to offer you. Agents who load you in the car and whisk you away to a few of their listings without conversation about what you're looking for are not serving themselves or you very well.

Experienced agents know that buyers don't always know exactly what they want. Time and time again prospective buyers will come into an office describing what they want only to eventually make an offer on something completely different. Buyers do not deliberately mislead agents. But choosing a home is a series of trade-offs and compromises with a good bit of emotion. A spacious family room with a picture window looking out on an exquisitely landscaped yard may make you forget that you absolutely have to be within three miles of the freeway. Therefore, good agents listen and observe intently, then read between the lines.

Agents will probably want to talk about your income and savings, which is unsettling because we don't talk about sex, death, and money, at least not with strangers. Obviously, agents have an interest in determining that you are financially qualified to buy a home within a given price range before they spend time showing you around. It is their responsibility to the seller. You don't have to be too specific and, taking into account issues of agency, you may not want to be if you're talking to a subagent of the seller. If you've read earlier chapters of this book, you have a fairly good idea of the size of a mortgage for which you can qualify. Tell the agent. Or if you're looking to buy a less expensive home than what qualifying ratios would allow, give the agent a price range and the assurance that qualification is no problem. Better yet, if you've prequalified for a mortgage, show the agent the documentation. The agent also may refer you to a lender for prequalification to avoid delving into your personal finances.

☞ **Money-$aving Tip #30** *If you want to play it close to the vest with a seller's agent, prequalify only for the size of mortgage you need in your price range, not the maximum you can afford.*

The second main area of qualification is your motivation. When do you need to move? When does your lease expire? Do you have children and want to time your move with the new school year? An agent may ask you a direct question such as if you were to find a house you like today, would you be prepared to make an offer on it. Agents want to determine if you're truly ready, and perhaps mean to challenge and prod you a bit. Agents know first-time buyers will be hesitant, that's human nature. It's hard to jump into the water when you've never swum before.

Agents will want to know the status of your home search. Have you been working with somebody else? How much looking around have you done and where? What do you think about what you've seen? They will want to identify the decision maker—the husband, wife, both, a trusted friend or

relative? The agent will want to get to know you, to understand your lifestyle and your tastes. Good agents will be asking
questions all the time as you go through homes, encouraging
you to open up, trying to determine what you like and don't
like. In general, the more you communicate with the agent the
better, keeping in mind where fiduciary duty lies.

You should be observing how closely the agent is listening.
If you're spending time visiting houses which clearly do not fit
your criteria, you might think about switching agents. Agents
may show you some homes which may not completely fit your
bill, but they should do so consciously. "There's a home you've
got to see—it's farther from the highway than what you'd like,
but the family room is unbelievable!" If you find yourself viewing homes listed by the agent's company and none other, you
have a problem. When selling its own listing, the brokerage
avoids splitting the commission with another company. If
you're looking only at one company's listings, you're missing
a big piece of the market.

You're the Boss

Whether your agent represents you or the seller, the goal is
to guide you toward your purchase. The agent will accomplish
this by showing you the right kinds of homes and developing
your confidence to take the plunge. You may very well need a
nudge here and there, and agents will be quick to oblige when
they see any sign of interest. The sales pitch is simply the
agent's job. There's no reason to be intimidated. Just sit back
and enjoy the art of selling, remembering that the customer is
king. You're the one bringing the money to the table. Use the
expertise of your salesperson to make the best possible homebuying decision. By reading up on the process, you are becoming a knowledgeable buyer, and the agent is your ally.

☞ **Money-$aving Tip #31** *Don't work with more than
one agent at the same time unless you truly have a compelling reason. A smart agent won't give you much attention if
you're working with others at the same time.*

How Your Agent Can Help

A real estate agent will alert you to homes on the market that fit your criteria and can give you printouts from the Multiple Listing Service. Tell the agent which homes you want to see and he or she will set up the appointments and show you the homes. If you'd like to make a bid, the agent will help you write out the agreement. The role the agent will play in the negotiations depends on the agency relationship.

All of the above is the bare-bones description of what the agent can do for you. However, agents can do much more. Truly, the agent can be your real estate consultant. With the caveat that agents (with few exceptions) are not lawyers and are not construction engineers, they nonetheless have loads of day-to-day experience that can especially benefit a first-timer. Below is just a partial list of the kind of support an agent might provide you:

- Background information on the neighborhood and its amenities, from schools and churches to grocery stores
- Plain language explanations of basic real estate principles and the many strange new words you're hearing
- Hard-to-get market information that is extremely localized and timely—such as the rate of appreciation of homes and how fast they are selling in an area
- Advice on financing strategies, potential lenders, and special community-based programs for first-time buyers
- Recommendations of contractors and tradespeople so you can factor anticipated repairs and renovations into your purchase offer
- Follow through with lawyers, escrow officers, lenders, and title companies to ensure a smooth close to the sale
- Moral support when nerves get frayed

Commonly Asked Questions

Q. What if I'm using a buyer's agent and want to make a bid on a for sale by owner (FSBO)?

A. It depends. You should let your buyer's agent approach the sellers. Most FSBO sellers are motivated by a desire to save the commission. Even if you're using a seller's agent, the homeseller will be reluctant to pay a commission. You can expect that they will be particularly reluctant to pay a commission to a buyer's agent. The alternative of dealing directly with the sellers isn't any better. Buying from a FSBO is a lot more work. You'll want additional input from an attorney for sure. You're probably better off focusing your search on listed homes, particularly as a first-time buyer.

Q. I've heard that I don't need a lawyer; is that true?

A. The procedure for the "close" of a real estate transaction (referred to as escrow in some parts of the country) varies widely from one area to another. In some areas it is customary for an escrow officer or the title company to handle the close. It is true that there is no law requiring you to use a lawyer. However, you want a lawyer on your side to protect your interests. The sooner the lawyer gets involved the better. You should have legal advice on your purchase contract, details on transfer of title, and all the documents of the close.

Q. Is seeing an agent the only way to find out about house listings in my area?

A. It used to be that the MLS, which is controlled by the real estate boards, had an exclusive franchise on maintaining computer databases of houses for sale. That is no longer the case. More and more listing services are cropping up on the Internet and online services. Nonetheless, REALTORS® still are likely to have the most comprehensive selection of what's for sale in your area. Because of the new competition in listings, it is all the more important for real estate agents to offer you services beyond simply identifying homes currently on the market. .

Location and Neighborhoods

On the odd chance that you don't know the three most important factors in real estate appreciation, we'll repeat them here: location, location, location. Sure, it's a cliche, but it's also full of truth. Nearly all other shortcomings in a home can be corrected. You will want to avoid problems that require huge investments of time and money to overcome; but if you're stuck with one, at least it can be solved. In the case of location, however, there's not a whole heck of a lot you can do.

The neighborhood you choose to buy in will shape your lifestyle and be a big factor in your family's happiness. Your choice of location and neighborhood is just as important from an investment point of view. Such factors as good schools, low crime rates, and convenience to transportation or amenities have everything to do with values in residential properties. In fact, one easy way to find the most desirable locations within an area is to simply look in the area with the most expensive homes. But most first-time homebuyers don't have the luxury of that approach. In this chapter, we'll identify the most important factors that contribute to a desirable location. You won't have to go to the most prestigious neighborhood in town in order to find them.

Larger Market Forces

The economic health of the metropolitan area is a big factor in home values. The same economic indicators that you hear about in the news will shape the values of homes in a neighborhood. Employment is the most important factor. If a lot of people are out of work, the value of homes will be down. The best areas to buy in boast stable and diverse employment opportunities. The dangers of a "one-company" town are clear. The most glaring example of an area dependent on a single industry is Houston during the oil slump of the 1980s. Home after home fell to foreclosure like dominoes, causing huge drops in values throughout the market. The same kind of effect can occur on a microeconomic level. In some suburbs, a large proportion of residents in a neighborhood or condominium development may work for one large employer. If that employer starts trimming its staff, home values in that area will go down. More homes will go up for sale in the market as people relocate to other jobs, and you won't have an influx of new employees looking to buy there.

You also should take a look at government and municipal services in an area. A well-run administration makes for a solid neighborhood: your tax money is spent wisely and efficiently; building codes are strictly enforced; zoning regulations are thoughtfully designed and adhered to; the garbage gets picked up; police and fire departments are there when you need them. You should have a clear idea of what your taxes will pay for. Sometimes unincorporated areas will hit you with special fees for services you might be taking for granted, such as garbage pickup or street cleaning. You'll want to know where the community gets its water supply and whether sewage disposal is through a community system or septic tanks that are the owner's responsibility.

Neighborhood Cycles

Neighborhoods define themselves by common characteristics such as housing of similar age, value, or style. Architectural styles will usually be similar, or at least complementary. The people within the community share much in common, including similar socioeconomic status, types of occupations, or values. In sum, there will be a sense of unity. Neighborhoods may have clearly defined boundaries, such as a river or railroad tracks, but not necessarily. Generally, you will be able to sense the change from one neighborhood to another.

All neighborhoods undergo a cycle of growth and decline. It's inevitable. At a given moment, the resurgence of an impoverished neighborhood or the decline of a prestigious one may seem inconceivable. The cycle may take a lifetime or more to complete, but it will happen. Neighborhoods will grow in value when new people come in and builders develop the land. Eventually housing prices will stabilize, a period that may continue for 50 years. But at some point the neighborhood will lose favor with homebuyers, perhaps because prices are too high. As people move away, the neighborhood will change. Large homes may be broken into apartments or be converted to commercial use. Then at some point that trend will reverse itself. You can see this happen in large cities. Artists, for instance, may flood into an area, attracted by low rents and commercial properties easily converted to working studios. Next come coffee houses, trendy restaurants, and art galleries. People visit the neighborhood for these amenities, notice the affordable housing, and start moving in. Developers catch wind of the neighborhood renaissance. Quicker than you can say condominium, those rough, unfinished artist's studios are converted to luxury lofts. This speedy reversal, called gentrification, though not uncommon in recent years, from a larger perspective is still not the norm. You might think of it as a time-lapse photographic look at a typical neighborhood cycle.

In the decades following World War II, middle-class families fled the cities in search of a better life in the suburbs, away from the social problems of large cities. Housing values in the cities dropped. Once grand homes and apartment buildings fell into disrepair because the owners were absent or financially unable to properly maintain them. Then the 1980s saw many of the sons and daughters return to the city to rehab these old homes. In some parts, a new phenomenon of suburban decay is beginning to unfold. For the suburbs closer to the city, building stock is now 40 or 50 years old. Not all the builders took a long-haul view, so some of the housing is showing its age. The 1960s saw a huge increase of apartment construction, much of it now in the suburbs. To many renters, the city is a more desirable place to live, and suburban landlords may have a hard time attracting quality tenants. Social problems like crime and gangs are now becoming a concern of the suburbs as well.

As a homebuyer, you want to be able to identify the neighborhoods that are stable or on the upswing. You should look for population growth during the past five to ten years. Chambers of commerce can supply this information. You also might check into the occupations of the residents as a gauge of the neighborhood's economic health. Signs of decline include land that is not being developed, homes in bad repair or in foreclosure, or a preponderance of older residents with no young people moving in.

Schools

One of the most important factors in determining the value of homes in a neighborhood is the quality of schools. If you are a parent or planning to start a family, you don't need to be reminded of this fact. However, it is just as important to the homebuyers without children insofar as it affects the investment value of the house. Without good schools, young people will not be attracted to an area. Residents will put their homes up for sale as kids reach school age.

Real estate agents can give you information about the schools in the neighborhood. Schools are so important that firms are cropping up to provide homebuyers and parents with information. One such company is SchoolMatch (800-724-6651), which for about $50 will give you a "report card" for any school system or accredited private school nationwide or, for a larger sum, will provide a customized report on the top schools in an area. Other companies keep similar databases on schools in specific areas. Test scores for schools within a system are sometimes published in local papers. Ask your agent or visit the library for other sources of local information.

☞ **Money-$aving Tip #32** *Low property taxes in an area seem attractive. However, they are also a sign of poor schools. The bulk of funding for most school systems comes from property taxes.*

Parents may want to take the time to tour a school. There's no replacement for a firsthand look. You can get at least a sense of the quality of the school by touring the grounds. A dilapidated building is a sign of a school short of funds.

Crime

We all want to feel secure in our homes and neighborhoods. Crime is obviously a significant factor in home values. If you live in a rural area or small town, you may not view this as a concern at all. If you're in a city, you are no doubt already attuned to safe neighborhoods and areas to steer clear of. Unfortunately, prejudices make many people consider the presence of minorities to be evidence enough of a crime problem. It's not that simple, though. Nor is it true that in leaving the city and heading for the suburbs, you are necessarily leaving crime behind you. The only reliable way of checking the level of crime in a community is by looking at statistics. You can contact local police districts or chambers of commerce for this information. The numbers are usually broken down into categories so you can distinguish between violent crime and property theft. A knowledgeable real estate agent should be able to access this information.

Families, of course, are quite sensitive to issues of crime. Other buyers, call them urban pioneers, may be willing to live in a marginal area. Why would you do that? You may want to rehab an old home into its former splendor, with hopes of turning a profit. Or perhaps you appreciate the amenities of city living, being close to cultural activities or employment, but can't afford the premium neighborhoods. If you're willing to take that risk, keep in mind that as long as crime is high, your property value will be kept down. Look for a neighborhood that is on the upswing. Concerned residents organized into block clubs or watchdog groups can make a big difference in cleaning up crime, especially as police departments support their efforts with community policing. Persistent neighborhood pressure has been known to chase away gangs and drug dealers.

Traffic and Noise

Most agents agree that heavy traffic is one of the biggest turnoffs to buyers. You should avoid buying on or very near a busy street. Heavy traffic is the leading cause of noise problems. Exhaust fumes are unpleasant and may pose a real health risk for people with respiratory problems. Cars present other potential safety problems, particularly for families with children. Major highways can cause noise several blocks away, although sound barriers along the highway can sometimes alleviate this problem.

In many older urban areas, streets are laid out in grid fashion. Side streets typically become natural spillover routes for the drivers trying to beat traffic. Curving streets and dead ends can make a big difference in reducing traffic through a neighborhood.

You should be on the lookout for other potential sources of noise. Close proximity to a hospital or fire station will mean wailing sirens. A house across the street from a school may be convenient for your kids and offer a place to play during off hours, but during the school year expect lots of noise during recesses as well as before and after school. You also may have to endure the fumes of a long line of idling buses. Other types of commercial establishments will cause noise or attract traffic. If you're located a block down from the stadium, the roar of the crowd will soon stir you in a different way.

Condominium developments tend to be high-density communities, so buyers should pay particular attention to noise. Being located next to a main entrance or near a parking lot will cause noise. Another major source of noise is one you can't naturally anticipate—your neighbors. A good idea when you're serious about buying a condo unit is to make several visits at different times of the day. If the resident upstairs has a taste for heavy metal at high decibels, you'll find out.

Neighborhood Amenities

What's considered an amenity may be largely a matter of taste or a function of communities. In dense urban areas, where parking spaces are scarce and traffic heavy, shopping that is walking distance is a real selling point. Proximity to public transportation also will be valued. Out in the sprawling suburbs, how quickly you can get on the freeway is more the concern. If you're walking distance to a commuter train line to the city center, that's a plus too. Many amenities are actually double-edged swords. You want to be close, but not too close. It's nice if shopping is nearby, but you don't want the mall adjacent to your property.

Recreational facilities can make a home much more attractive to you personally, but may have little impact on value. For instance, it may mean a lot to you that a house is just a mile from your favorite bike trail in the area and you won't need to ride on any busy streets to get there. If others hardly notice, then this is one amenity that won't cost you.

Getting to Know a Neighborhood

In addition to formal research, the best way to get to know an area is to drive around at all hours, attend open houses, and ask real estate agents and residents lots of questions. Better yet, get out of the car and walk around in a neighborhood. (Riding a bicycle is good too, and you can cover more ground.) General observation will tell you a lot, and you won't see (or hear or smell) nearly as much cruising by in a car.

How does the neighborhood look? People make all the difference in a neighborhood. Do you feel a sense of pride among the homeowners? Are the houses well maintained, the grounds neatly kept? Look for every opportunity to strike up conversations with residents. Ask them about the neighborhood. After a while, a neighborhood becomes a part of who you are, and most people welcome the opportunity to talk about it.

These walks might make characteristics of the neighborhood you would otherwise miss readily apparent. Smoke and odors may come from some distance to affect a neighborhood. You may even learn that the neighborhood is on a flight pattern to the airport. Make your visits at different times of the day. You may see loads of kids out playing on the street. Or teenagers recklessly racing down the streets when the nearby high school lets out. Try out your commute to work from the neighborhood and the return trip too. Will you be able to manage it every day?

Facilities or elements that tend to decrease home value if located too close to the property include heavy industry, vacant houses, railroad tracks, motels, power lines, apartment buildings, bars, billboards, cemeteries, funeral homes, fire stations, and hospitals.

☞ **Money-$aving Tip #33** *A real estate agent always will take you to a home by the most pleasant and attractive route. Do some looking around yourself to see what the surroundings are really like.*

Many communities publish neighborhood newspapers. Spending an afternoon at the library browsing through these is another way to pick up the flavor of the neighborhood. If you're set on a neighborhood, you may want to subscribe. It's a good way to learn about plans for future construction or development that could impact property values. Don't overlook maps either. The bird's-eye view they offer, patterns of parks, highways, or railroads, can be quite revealing.

Zoning

The birth of zoning regulations came in the early 1900s with the goal of controlling rapidly increasing population density, traffic congestion of the cities, and the growth of commercial structures in residential neighborhoods. Since World War II, the emphasis of zoning has been to control land development and design to prevent the spread of urban congestion and poor building design. People sometimes try to use zoning to stop development altogether. Its best purpose, however, is to control development so schools and other key municipal services may expand in an orderly fashion.

You should understand the zoning regulations that apply to your intended area. What are the rules concerning commercial establishments? Is the development of multifamily housing, either apartments or condominiums, allowed? You should be particularly attentive to this detail if there is undeveloped land near a home you're considering.

Contact the local zoning board to learn about regulations. You also should ask about variances, which are exceptions granted to residents for certain construction. Zoning boards have a history of generously granting variances. As an example of how this might affect you, a new next-door neighbor may have been granted a variance to construct a large addition that has a detrimental effect on your backyard view. You also should contact your local public works department to learn about upcoming projects. Agents can help you with this information.

You also should look into CC&Rs—conditions, covenants, and restrictions—that are on the public record. CC&Rs may dictate such details as what color you can paint your house. They are meant to protect the home values in an area, but nonetheless can be restrictive.

Site and House Location

The preferred physical characteristics of a neighborhood are paved, winding streets, shade trees, and attractive views. Most people prefer rolling, hilly terrain with the neighborhood on ground higher than the surrounding area, over a flat or rugged, mountainous terrain. Understandably, areas near marshes or stagnant ponds, with poor surface drainage and at risk of flooding, are to be avoided. The same goes for the topography of the house site. Look for grading that will drain water away from the home.

The location of the site and the orientation of the home on it are also important, though they have often been overlooked by planners and developers. The tendency has been to line up houses one next to another, all facing the street, when alternative methods might be preferable. The land in front of the house is called the public zone, the back the private zone. Some frontyards are expansive with the homes set toward the back of the lot. Unfortunately, this means lots of maintenance, cutting the lawn, or shoveling snow from a long driveway at the expense of the recreational use of the land. Backyards are preferable for children or entertaining guests outdoors. For this reason, corner lots, which may seem attractive, actually are not. You'll have a large, wrap-around frontyard and a tiny backyard, or none at all.

You may come across irregular lots such as *key lots* or *flag lots.* In some tracts, a key lot is adjacent to a corner lot. Instead of a sideyard, it butts up to somebody else's backyard. A flag lot is shaped like a flag. It comes to be when a developer ends up with an irregular tract of land in the center of the development. To squeeze another home and more profit out of the land, the developer runs a long access driveway between two properties and builds a home which is surrounded on all sides by the backs of other properties. You'll get a home on these types of lots for a low price; but you'll have to sell it at a low price too, and it may take a while to do so.

When assessing the house's orientation on the lot, give some thought to the movement of the sun. In the heat of summer, the sun rises in the northeast and travels in a high arc toward the northwest. In the cold winter months, the sun will rise in the southeast, traveling across the sky in a low arc to set in the southwest. You'll get most of your sunlight on the south side of the house. Ignoring other considerations in the surrounding area, the best orientation for the house would be with a wide side facing south and containing large windows. A roof overhang will serve to shade the house in the summer months while letting in warming rays of the sun in the winter months, when it is lower in the sky.

Consider also privacy when looking at the orientation of the house on the site. Large picture windows close to the street create a fishbowl effect. You may find yourself closing the drapes often, which is not nearly as nice as enjoying a pleasant view.

Good landscaping not only adds beauty to the property but it also adds functional benefits. Shade trees keep the house cool in summer months, and shrubs can be great for privacy.

Orientation can make a big difference in condominium and town house developments. Obviously, units overlooking the well-manicured common grounds command a higher price and will be easier to sell than those with a view of the parking lot or a noisy highway.

How *Your Agent Can Help*

We encourage you to have a general idea of where you'd like to live before you settle on an agent. This advice is particularly true in large metropolitan areas, where you can't expect an agent to know much about a neighborhood clear across town. In small towns, agents may be quite knowledgeable across all the communities. Once you've narrowed your choice to a general vicinity, the real estate agent can be a tremendous source of information about all the factors discussed in this chapter.

You should start by making a list of the factors most important to you in a neighborhood. The factors may be general, such as good schools and convenient transportation, or perhaps more specific. Maybe it's proximity to a freeway or train line you take to work. Maybe it's a particular school district. Add to that another list of features you'd like to see in the home: number of bedrooms, particular housing styles, and the size of the yard. Then let the agent know your price range.

At this point, you may be ready for a reality check. Often, first-time buyers learn that they must compromise on some aspects of their dream home. Thus the term *starter home*. In any event, the agent will then have enough information to point you in the right direction. You should check out neighborhoods on your own. The agent's focus will be showing you houses. Nonetheless, agents pick up plenty of neighborhood expertise in the course of their work, so don't hesitate to ask questions.

Commonly Asked Questions

Q. *How can I learn about a neighborhood when I'm relocating to a new city?*

A. You'll have to learn about neighborhoods the same way as described in this chapter. As a practical matter, you most likely will not have time for some of the checking around suggested here. Logistics certainly discourage it. You can learn a bit by subscribing to the local newspaper, and you can contact the city's chamber of commerce as well. Also, the agent becomes a more important resource. For this reason, relocating buyers are good candidates for buyers' agents.

Q. *I've found a neighborhood I like, but homes there rarely go on the market. How do I get one?*

A. First, you'll have to be patient. If you have a strict timetable, you better start looking at other neighborhoods. Second, you should work with an agent from a company that is the leader in listing homes in the area. Ask your agent to notify you as soon as a home fitting your criteria goes on the market. Third, you need to be prepared to act fast with a good offer. If the area is so desirable, you are not alone. Preapproval (not merely prequalification) from a lender would allow you to forgo the mortgage contingency and make a more attractive offer.

One enterprising idea is to distribute flyers around the neighborhood telling homeowners about you and your family and letting them know you're looking to buy. While certainly unconventional, this technique may be worth a shot. An owner considering a sale may seize the opportunity.

Q. *How do I know if a house is a flood risk?*

A. The surrounding area of swamps or marsh could give you a strong indication. You also can look for signs of water damage in the house. Beware of fresh coats of paint or paneling in the basement. Seller disclosure laws, which most states have adopted, will require owners to disclose a flooding problem. You also could contact the closest office of the U.S. Army Corps of Engineers, which prepares maps of flood plains for the Federal Emergency Management Agency. Lending institutions have maps, too. Your agent can help you locate this information.

CHAPTER 8

The Well-Designed House

You'll save yourself a lot of frustration if you pay close attention to design when you're looking at houses. You are nearly as helpless in correcting design flaws as poor location. While these problems often do have a remedy, they may require extreme and expensive measures. Flaws like a zigzag, convoluted path from the kitchen to the front door may seem slight and are easy to overlook when you're viewing homes. Living with such a problem for ten or so years can drive you nuts.

Solid home design consists of three components:

1. A well-planned orientation of the house to the site
2. Attractive exterior appearance
3. Good interior planning

The first of these factors was addressed in the previous chapter. We will look at the second two in this chapter.

Don't confuse interior planning with interior design, which has to do with decorating along with the selection and placement of furniture. It's a proven fact that tasteful decorating will help sell the home. Sure, homebuyers know that the decorating, with the exception of fixtures, will be gone when the owners leave the house. Nevertheless, an attractive home has a subconscious effect on a viewer. You just want to like it. Even if the buyer would do things differently, the clear message is "this place can really look nice." In other homes, that message may be just as apt, but you'll have to use your imagination to see it. Don't overlook the owner's decoration altogether. The design of the home itself sets certain parameters, and you can observe how one person chose to work with them. If your taste is for contemporary design, an old house with wood trim and antique details is not right for you. The bottom line is that nine times out of ten you'll want to do some redecorating when you buy a home. Decorating can be expensive, so you can't ignore the cost. On the other hand, it still should be a relatively minor factor in your choice of homes.

The interior planning and exterior appearance of the house are givens. It's best to buy what you want rather than to make modifications to a home that doesn't fit your needs or criteria.

Families or couples planning to have children should pay special attention to design features that will make raising the kids easier. It's nice to have a clear view of young children's play areas from the kitchen. Stairs are always a potential hazard. However, keep in mind that children don't stay small very long. They will learn stairs. Bedrooms close together are ideal for young children, but teenagers would prefer to be as far as possible from their parents. You'll feel the same way about them.

Attractive Exteriors

Houses and buildings are living pieces of art, and the great architects will always debate the interplay of form and function. Like a work of art, it's difficult to describe the many details that contribute to a beautiful house. Most of us nonarchitects can only say we know what we like. More precisely, when good design practice is absent, it's glaringly obvious to even the most casual of observers. We take great pride in our homes. We'd all like the exterior to please the eyes of friends and relatives when they come for a visit.

Here are a few things to look for:

- Houses should show a unity of style and building materials. Mixing a couple of materials is fine if done tastefully. Any more than two building material types is probably too much.
- The doors and the windows of the house should line up and be in proportion to each other and the house. A grand entrance may be perfect for an antebellum mansion but will look silly and pretentious in a three-bedroom suburban home.
- Simplicity and restraint nearly always win out in house design. Excessive ornamentation, often added to give a sense of luxury to an otherwise modest home, gets tacky fast.
- Observe the lines of the roofs and walls. Avoid lots of corners and zigzag walls that make the house look busy.
- Look for a sense of visual organization and balance of proportion. Parts of the house should relate to one another.

Architectural styles tend to change over time. Early in this century the Cape Cod and bungalow designs enjoyed immense popularity. They are not built much anymore. Colonial and Georgian style two-story homes have never gone out of fashion. After World War II, the ranch home spread from its beginnings in the West across the country. It's widely accepted and appears to be here to stay. Split-level homes were first designed to address the concerns of a graded lot, but new varieties have since been developed for placement on flat lots. Choice of architectural style is mainly a matter of taste and lifestyle. What's important is that the style is well executed in the design, both inside and out.

Number of Floors

We will look at four varieties of house types based on the number of floors.

One-Story Houses

The most obvious advantage to the one-story house is that there are no stairs to climb, a critical factor for the elderly or people with disabilities. Parents also won't have to worry about young children taking a tumble. Exterior maintenance is easier. A short ladder will get you up to the roof to clean the gutters. The ranch home is the most common variation of the one-story home built in recent years. It's not surprising that this design had its beginnings in California. One-story homes present excellent opportunities for integrating indoor and outdoor living.

The one-story design, of course, does not use land as efficiently. It needs to be spread out to deliver living space comparable to a two-story house. Good interior design is critical. It may be difficult to separate bedrooms from the active areas of the home. The noise of parents entertaining, for example, will disturb children trying to sleep or study in their rooms.

One and One-Half Story Houses

One and one-half story houses are not built much anymore because they don't cost more to build than two-story homes. Still, this variety has a big presence among existing homes. They deliver a large amount of living space in proportion to the size of the structure, leading also to efficiencies in heating and air-conditioning. However, a common problem is that there is insufficient insulation under the roof and poor ventilation in the second floor, so be careful to check the heating and cooling of the second floor. These rooms are often hot in the summer and cool in the winter. They may also be cramped and lack windows. You can solve this problem with dormer windows, but they are an expensive addition.

Two-Story Houses

The two-story house is the American classic. Tradition is on its side, along with a feeling of gracious living. The problem of separating bedrooms from the living area solves itself. Bedrooms upstairs provide privacy and often a sense of security. A style of two stories (or more) provides maximum living space when lot sizes are small. The disadvantage, of course, is the darn stairs, which besides taking energy to climb also use valuable space in the home. A common design flaw is poor choice of stair location.

Bilevels, Splitlevels, or Raised Ranches

Slight nuances differentiate these houses. The common denominator is that different parts of the living areas are placed at separate levels and connected by short stairways. At its best, the splitlevel combines the desirable features of one- and two-story houses. The sleeping area is easily separated. In its original form, designed to a sloping lot, the plan provides for access to the outdoors on two separate floors.

You'll basically need more stairs. Imagine yourself moving through areas of the house and consider what your most common paths would be. Would you be constantly going up and down stairs? This may or may not be the case in split- levels. You should take a good hard look at design if the house is on a flat lot.

Basements of splitlevels are often raised up out of the ground several feet, providing greater window depth and more light, and thus a usable living space. However, care must be taken that the basement is dry and properly insulated. It may be difficult to keep the basement at heat levels consistent with the rest of the house. Heating and insulation of the highest parts of the house are also concerns.

☞ **Money-$aving Tip #34** *Good design of a small, splitlevel house is difficult. You'll see better results in homes of 1,500 square feet or more.*

Zones in the House

When judging the floor plan of a house, you should think of it in terms of zones. The house is commonly divided into three zones:

1. The *living zone* includes the dining room, living room, family room or den, and enclosed porch.
2. The *private* or *sleeping zone* consists of bedrooms and bathrooms.

3. The *working zone* traditionally consists of the kitchen, pantry, laundry facilities, and workshop. You also might add home office to this list, though spare bedrooms often double as offices.

A good floor plan separates these zones as much as possible so that activities in one zone do not interfere with those of another. Hallways and staircases allow movement between zones while creating a sense of separation. The house will have a guest entrance, commonly in front, and one or two others, called family or service entrances, at the side or back.

General principles define a design that properly separates zones and still provides good circulation through the house. The main entrance ideally should lead into a foyer or hallway, preferably with a closet. The natural pathway should be toward the living room, though you don't want the doorway opening directly into the room.

Ideally, hallways should be configured so you can move from one room to another without entering a third. The living room, particularly, should be a dead-end room. It is a place to relax and converse without being disturbed by kids running in and out.

The kitchen is command central of the house. Ideally, it should be located where other parts of the house are easily accessed. One of the most common paths in the home is from the kitchen to the front door, so it should be a clear shot. You'll want the eating area or dining room directly adjacent to the kitchen, so you can bring food to the table without walking through a hallway. You also should have direct access to the outdoors through the service entrance. Patios or decks should be easily accessible through this or another entrance.

Bedrooms and bathrooms should have the greatest privacy. Preferably a barrier should block sound and sight between the sleeping and living zones. You should be able to pass from any bedroom to a bathroom without being seen from the living room.

Where design is good, it usually goes unnoticed. Give some thought to the floor plan where you live as well as to plans of houses you frequently visit. Chances are that your favorite houses have superior floor plans.

Common Floor Plan Flaws

Here is a list of design flaws you're most likely to see in houses. Do your best to avoid them.

- Front door opening directly into the living room
- Obstructed pathway between the kitchen and front door
- Service entrance to the kitchen too far from the driveway or garage
- Bedroom or bathroom clearly visible from the living room or foyer
- No comfortable eating area in or near the kitchen
- Formal dining room too far from the kitchen
- Poorly located family room
- Too many doors or windows in a room, making it difficult to place furniture
- A front door with no window or opening through which to identify visitors
- Gas, electric, or water meters located inside the house so you must let in meter readers, a security concern
- An oversized picture window at the front of the house, sacrificing all privacy
- No light or electrical outlet on patio or deck
- Doorways that open directly to basement stairs, inviting accidents

The Kitchen

The kitchen is perhaps the most important room in the house. A well-designed kitchen will always boost the value of a house. As discussed above, the floor plan should recognize the room's prominence with a central location. In this section we'll look inside the kitchen.

Modern needs in kitchens have changed quite a bit in the last generation. We need more counter space to accommodate new appliances like food processors, microwaves, toaster ovens, juicers, and coffee grinders. With today's lifestyles, it's not unusual to have two cooks working in the kitchen at the same time. In some new homes kitchen designs include two sinks and separate work areas. Storage in the kitchen has always been at a premium. Now that we're all recycling, we need a convenient place to store our empty cans and bottles as well.

Efficiency is the operative word in kitchens. A kitchen door should swing out of the kitchen. If not, it should swing toward a cabinet or away from hallway paths or activity areas. Work areas should be separated from hallways and access points. Look for easy movement within and in and out of the kitchen.

The four main work areas of the kitchen are (1) the refrigerator; (2) the sink; (3) the food preparation area; and (4) the oven. Ideally, cabinet space should be available in different parts of the kitchen so items can be stored near where they'll be used. Tall appliances, such as a range, should not break the flow of counter space. Food preparation areas should be placed between both the sink and refrigerator and the sink and the range. It's best to have counter space on both sides of the range so pot and pan handles don't stick out into pathways, inviting spills and accidents.

The area between the sink, the refrigerator, and the range is called the work triangle. Believe it or not, a kitchen can be too big. The three sides of the work triangle should total no more than 26 feet. You don't want any wasted steps!

Bathrooms

The biggest question with bathrooms is how many. Years ago, families of seven people or even more would somehow find a way to share one bathroom. Today, one and one-half baths are considered the minimum, with two or three preferable in a home of three bedrooms or more. A half-bath refers to a lavatory—toilet and sink only.

The largest bedroom, called the master bedroom, often has its own bath. This isn't necessarily good design, at least not for families in homes with only two baths. Guests will have to use the children's bath, which tends to get dirty and cluttered with bath toys and other kids' stuff. Watch out for baths with two entrances, one from a bedroom and the other from a hallway. Baths also might be placed between bedrooms with two doors. This makes for a lot of confusion and can disturb those in the bedroom with noise or light from under the door. Families should be sure that bathrooms are big enough to allow an adult and two children to move about freely at the same time.

☞ **Money-$aving Tip #35** *Homes with only one bath or fewer than three bedrooms will be hard to resell.*

Family Room

A family room is a relatively new entry into home plans. It is a byproduct of the postwar baby boom and the accompanying suburban housing boom. As families grew, parents sought a place for their kids to play. Where that room was located depended primarily on what space was available in the house. In many cases, basements were finished with paneling and floor covering. As the concept grew in popularity, the family room moved closer to the main living areas. Interestingly, the terms recreation room or den were once used for this space. Now family room is the more common moniker, an expression of how everyone has embraced this space.

The emergence of the family room has changed the role of the living room. Previously, good design had dictated that the living room be positioned for the best view in the house. It is now considered appropriate for the family room to occupy this position. For many years, the living room was where you'd find the television. Now the best television is often in the family room with smaller sets in the bedrooms. As in the past, the living room is reserved for conversation with guests.

Other Areas of the Home

Number of bedrooms is also a key factor in home selection. Three bedrooms is probably the ideal size. Families with two children can give each a separate room, and larger families can separate the sexes. You can always squeeze the kids together to make room for a guest. Two bedrooms might work for you, but selling could be difficult. A four-bedroom house is nice and may give you a spare room for guests or an office. But four bedrooms also could be a liability when it comes time to sell. You will be appealing to a limited pool of buyers. Today's smaller families no longer require four bedrooms, so it is becoming a luxury feature. Looking ahead to the sale of your four-bedroom home, the law of supply and demand may be working against you.

Bedrooms need only be big enough to hold the bed and furniture, and provide room for dressing and cleaning. Some homes feature bedrooms expanded into multipurpose suites. The design should be well thought out. Otherwise, these rooms simply waste space that would be better used elsewhere, in living zones.

The latest room that's becoming standard is the home office. Only the newest of houses will feature rooms built specifically for this purpose. Extra bedrooms often are converted to offices. That means the office will be in the sleep zone, which may work out okay. However, a location in the work zone is more appropriate.

☞ **Money-$aving Tip #36** *Be sure to consider electricity when you consider rooms to be your home office. You may have to add a new circuit or additional outlets to run your equipment.*

Laundry equipment can be placed anywhere where plumbing and gas or electric hookups are available. It's largely a matter of personal preference. Some large homes feature utility rooms. The basement can be an ideal place for the washer and dryer, although some would rather avoid the trips up and down the stairs.

Garages, carports, and driveways should allow for convenient movement into the house. Attached garages have obvious benefits, especially when the weather gets messy. The two-car family is so typical that the two-car garage might be considered a necessity, though in the Sun Belt it's not as critical. Whatever the climate, it's usually easy to choose one of the two cars to keep outside.

Patios, terraces, and decks should be designed for easy movement in and out of the house. We Americans love cookouts, so access to the kitchen is important. An outlet is convenient for yard work with electric tools, and light fixtures can enhance your outdoor entertainment in the evening.

Last, but not least, look for closets and storage. Whatever you have, you'll use.

How Your Agent Can Help

By definition, neighborhoods consist of similar or at least consistent house styles. If you relate the home of your dreams to a specific style, be sure to let the agent know. The agent can tell you where in town you'll find it. If the answer is, "It's not done here," discuss what it is you like in the style. The agent may be able to suggest something else that fits the bill.

The same goes for virtually any design feature that appeals to you. For example, as mentioned earlier, most houses are oriented front-to-the-street in cookie-cutter fashion. There might be exceptions in your area. If you'd really like a different house orientation on the site, the agent can tell you where to find it. Whatever features are important to you, let the agent know.

Commonly Asked Questions

Q. What about knocking down interior walls to create new spaces?

A. That can be a great idea, or it can be a disaster. You need to consult a structural engineer or contracting expert. Do so before you buy the property, of course, if the change is critical to your decision.

Q. Landscaping changes would vastly improve my site. What can I expect to pay?

A. As a ballpark figure, experts estimate that new landscaping will cost about 10 percent of the value of the home. You may want to call in a landscaper for an estimate before you buy the property. Though the money is significant, landscape improvements are one of the best for return on your dollar. If done thoughtfully, you may bring down energy costs as well.

Q. How hard is it to add a bathroom to a house?

A. If you can work where plumbing lines have been roughed in, it may not be so hard. Your home inspector may be able to advise. Here again it would be wise to bring in a contracting specialist before you buy. This might be a good fixer-upper strategy for a one-bathroom house, but be certain of what your costs will be. Under the right circumstances, you can get your money back and more.

CHAPTER 9

Strategies for the Cash-Strapped Buyer

No doubt about it, homebuying will never be as tough as the first time around. You are probably amazed and aghast at home prices. If you believe they are just too expensive, that's only because they are. In the 1980s, buying a home was like chasing a carrot on a stick. Fast as would-be buyers could save, they still couldn't keep pace with rising home prices. The goal was simply to "get in." That has always been the challenge and will continue to be. Once you're in, you too benefit from appreciation and have that much more to spend on your next purchase. The real estate slump of the early 1990s stabilized prices in most parts of the country and brought them down in a few. But buying a home is still a dream you will probably have to sacrifice for.

Fortunately, government and society value home ownership. The more people who own homes the better it is for the economy. The lending industry as well as government agencies therefore have devised programs to help make homebuying more affordable.

Surveys show that the biggest obstacle for first-time homebuyers is scraping up the down payment. In this chapter, we'll look at several ways you can get into a home for less. The right financing makes all the difference. The FHA and VA loans are long-standing programs from government agencies that help buyers borrow more and put less money down. Those programs will be described in detail in Chapter 10. Here we will look at some creative financing techniques, such as lease options, equity sharing, and seller financing. If you're up late channel surfing, you may have come across cable TV hucksters, selling the dream of no-money-down riches on foreclosures. Foreclosure riches are possible, but are really more for the sophisticated investor who does have cash. However, we will look at a few realistic alternatives for you in the foreclosure market.

The first idea that comes to mind for homebuyers wanting to save is to buy a fixer-upper. You won't necessarily be able to put less money down, but it's possible you can get a home in a desirable area for less. But fixer-uppers is another area where a reality check may be in order. All too often, the "handyman's dream" becomes a nightmare. You'll have to do some good planning up front to make the fixer-upper strategy work.

The common thread that unites all the strategies in this chapter is that they require an incredible amount of work. Not surprisingly, if you want to save money, you'll have to put lots of research time into it. Don't take this lightly, because you'll be taking on more risk as well. For any of these strategies, you'll also have to rely heavily on the sound advice from your team of experts—contractors, lawyers, and accountants. Don't go it alone! Mistakes can be costly. The purpose of this chapter is to identify possible strategies. If any one of them is of serious interest, it is highly recommended that you research it thoroughly and enlist the help of experts.

The Fixer-Upper

Call it the *This Old House* phenomenon. You buy a nice old house in an established neighborhood, easily accessible to work and the cultural attractions in town. The idea is simple, so alluring. Find an old house. Save thousands buying it. You love home projects. With care and craftsmanship, you fix it up into the home of your dreams. It means so much, because you've done the work. You may choose to sell right away and make a tidy profit, maybe do it all over in another house. More likely, you'll want to stay a while and enjoy your accomplishment. So goes the dream.

The more common reality goes something like this: Get what you think is a good deal on the house. (It's cheaper than anything else in the neighborhood!) Drive yourself crazy for two, three years trying to fix it up, growing weary as every spare moment is spent working on the house. Struggle to keep your marriage together as loving gazes become whose-idea-was-this stares. Bore your friends with one horror story after another about working with contractors. Sign check after check. Look for more financing because the work is not done and you've run out of money. At last, the home of your dreams is complete. Enjoy. Enjoy for a long time, because you'll have to stay a while if you ever expect to get your money back at the sale.

You can find fixer-uppers in any neighborhood. In some neighborhoods virtually every home on the market is a fixer-upper. The problem is you won't want to live there. However, if you find a home you like in a marginal neighborhood that's fast improving, your chances of making it work economically are much better. You'll also find the handyperson's special in upscale neighborhoods. But not nearly as many, and you will most likely be in a competitive buying situation. A desirable neighborhood will always have many buyers competing at the low end.

A few other factors work against you in finding a good home for rehab in an upscale neighborhood. First, the land is valuable. No matter what the condition of the home, the location will prop up the price. Second, sellers are more likely to fix up the home before it goes on the market. They'll have the wherewithal to do it, and it will pay off because the neighborhood is good. They'll want to get what their neighbors are getting and will be willing to make cosmetic improvements like painting or new carpeting to get it. They may even remodel kitchens and baths. Pretty soon, it's not a fixer-upper anymore.

Your best bet is to find a home with a major problem or two that will cost tens of thousands of dollars to repair. Most owners will not spend that kind of money unless they're going to stay. You'll be looking for something that will scare off most buyers, a sagging structure for example. Then you'll need to be certain that the problem can be solved and what it will cost. Perhaps all that's needed are new supporting beams in the basement.

Is It Right for You?

To make this strategy work you better be sure you have the right skills and personality, as well as a lifestyle that permits investing many hours in home projects. You'll need the skills to do much of the work yourself if you expect to make any money. Tradespeople are expensive. If you're relying on substantial financing, you'll also have to convince the lender that you're capable of the work.

You will have to bring in professionals for some of the work. Expect to make a lot of phone calls trying to track them down. Then you'll have to prepare specific descriptions of the work you'd like done. You'll need to escort them through the house for estimates. You should be there at least some of the time they're working to make sure you're getting what you paid for. If you set up the job with a supervisor who's loose on the management side, the workers then come in and start improvising. You can imagine. "No, I wanted the south wall knocked down, not that one!!" Of course, you should inspect the finished job. In short, even having other people do the work will take time and money, and you will have to have the personality to manage people forcefully and organize work flow.

☞ **Money-$aving Tip #37** *Be sure to factor in your time when you estimate the costs of fixing up a property. Do so generously. It will take longer than you think. Consider what you'd otherwise be earning or what you'd have to pay somebody for the work and use the higher figure.*

Buy Right

The most common mistake people make in buying rehab properties is underestimating the costs of repairs. In the end, and after much aggravation, they have spent more than if they had bought a comparable home already in good condition.

Instead of focusing on what the owner is asking or how much less you will pay than what other homes sell for, concentrate on what you will need to spend to bring the home up to speed. You'll want to be able to make a fairly good estimate of costs before you look at the property. If at all possible, look at homes with somebody who has some experience with major home repairs. Your agent may be able to give you some advice, but don't take it as gospel. Agents look at a lot of houses, but most are not construction experts.

Ideally, you should bring in a structural engineer or general contractor to look at the property before you make a bid. The only reliable way to estimate costs would be to have various tradespeople review the property, but practically speaking, it's not possible. You also run the risk of a tradesperson becoming a competing buyer. People in the trades often invest in real estate.

Ask your agent for an estimate of what the home would sell for if it were in good condition. Subtract the estimated cost of repairs, including the value of your time, from the agent's selling price estimate. Now you have the maximum price you can pay for the property. Perhaps you'll want to factor in some profit for yourself as well. The owner may be insulted that your figure is so low, or may pretend to be. You may be in a competitive situation where other buyers are willing to go higher. Stick to your guns. Better to walk away than fall into one of the most common of homebuying traps: paying too much for a fixer-upper.

Financing a home you will rehab is usually a two- or three-step process. You get a mortgage loan to buy the house and a loan to fix it up. You can then refinance after the improvements have been made and get a loan covering the full value of the home. The FHA's 203(k) loan allows you to finance the purchase and the repairs with a single mortgage. Fannie Mae offers HIMLs (home improvement mortgage loans) and CHIMLs (community home improvement mortgage loans). Contact lenders for the details.

Foreclosures

The get-rich-quick deals are indeed out there, but don't bother looking for one. First, most of them aren't no-money down. On the contrary, it's likely that investors with ready cash to put down along with the ability to act fast will get these deals. To make a killing requires experience and some sophistication with real estate investing. The big money is in dealing directly with owners facing foreclosure, sometimes deceptively, or buying at auction "on the courthouse steps." Foreclosures, like any other investment, follow the risk-reward equation. Big money means big risk. Buying your first home isn't a time for high-risk investment. However, banks, savings and loan companies (S&Ls), and Fannie Mae do offer foreclosure properties that present a low to moderate risk for first-time buyers. You won't get any big price breaks. But as an owner-occupant, you will have good financing options. The homes also will be in reasonably good condition. Foreclosure properties also are available from the Department of Housing and Urban Development (HUD), which insures on FHA loans, and the Department of Veterans Affairs (VA), which administers the VA loan program. These properties are better suited to investors than first-time buyers. Procedures are complicated and involve intricate bidding processes.

Real Estate Owned (REO) Properties

A good option for first-timers is an REO, which stands for real estate owned. These are properties which the lender has taken over through foreclosure. They are the repos of the real estate world.

The chief benefit of REOs for the homebuyer is attractive financing. You may be able to get these properties for down payments as low as 5 percent or 3 percent, or even with nothing down. The bank also may be willing to relax underwriting standards. For instance, they may overlook past problems in your credit history. You're likely to get a good interest rate as well.

Banks usually have separate departments that handle REOs. The banks may sell the homes themselves, but more commonly they are listed with brokers. The purchase will proceed

similarly to standard properties, starting with the buyer's sub-
mission of a purchase contract. Usually the homes will be fixed
up before the sale. The brokers who list these properties are
sharp and experienced. They'll drive a hard bargain. Don't expect to get a price under market just because the property was
foreclosed upon. In fact, banks and S&Ls sometimes have un-
realistic expectations about what the home should sell for.
They may focus on the costs they need to recover, rather than
the realities of the marketplace. Dealing with the bureaucracy
of the lending institution slows the sales process and may lead
to some frustrations.

☞ **Money-$aving Tip #38** *Occasionally, entire blocks of
condominium units may be held by a bank or S&L and sold
at auction.*

A couple words of warning on these properties. Though
they are fixed up, an addendum is often added to the contract
stating that the house is sold on an "as is" basis. A thorough
home inspection before you purchase is quite important. You
will have no recourse after the purchase. You also should be
sure your attorney thoroughly reviews the purchase agreement and documents of the sale. Buyers will probably get a
special warranty deed, which does not provide much protection against title problems. Banks and S&Ls usually provide title insurance. The title policy may be subject to the Durrett
Rule exclusion, which means that the home could be lost in
the bankruptcy of the foreclosed-upon former owner. This is a
danger if the bank or S&L has owned the property for less than
a year.

REO properties that are listed with brokers will go on the
MLS. You can ask your agent to look for them. You also might
make calls to the major banks and S&Ls who make mortgage
loans in your area. Ask to be put on the mailing list to be informed of REO properties. Not all banks and S&Ls will offer
this service, but some will.

Fannie Mae

Fannie Mae is a secondary market purchaser of mortgages. They may hold a mortgage while a lending institution services the loan. If the buyer defaults and is foreclosed upon, Fannie Mae ends up with the property. Fannie Mae foreclosures are usually listed with brokers and also will be posted to the MLS, possibly with a message stating that special financing is available, or they may be sold at auction.

Fannie Mae does a good job of fixing up the properties. In fact, the biggest selling point of Fannie Mae properties is their condition, which is the best you'll see among foreclosures. Still, some critics argue that they only make cosmetic repairs. The houses are sold on an "as is" basis, though an option of inspection is available.

Fannie Mae foreclosure properties have much in common with REOs. The professional brokers who list their properties won't give anything away. Fannie Mae may offer some incredible financing on a house, but not necessarily. You will have to meet all of the standard underwriting requirements, no exceptions. You also will have to pay all the closing costs and fees associated with the loan.

Buying real estate at auction is not for the faint of heart. Fannie Mae auctions, however, are doable, even for the first-time homebuyer. They are glitzy affairs held at large hotels. There is no fee for attending. A broker can register you, or you can watch for ads and register directly. You have the right to inspect properties beforehand. You can usually get prequalified for a mortgage at the auction itself, so you'll know how much financing is available according to Fannie Mae rules.

Special Lender Programs

Mortgage loans require down payments of 20 percent of the purchase price. Anything less than that is considered to be too risky for the lender because the leverage is too high. The lender looks at its side of the equation through a measurement called loan-to-value ratio (LTV), which is simply the amount of the loan divided by the value of the property. Put another way, the LTV is 100 percent minus the percentage put down. The LTV on a loan with 20 percent down is therefore 80 percent. If all buyers had to put 20 percent down, the real estate market would slow down considerably. The remedy is mortgage insurance, which insures the bank, not the borrower. The insurance will back up the loan, making compensation to the bank if the borrower defaults. Like any insurance, it's all about covering risk. The VA and FHA programs do not actually make loans, they insure or guarantee them. Buyers who don't qualify for these programs can still put less than 20 percent down, and pay the premium for private mortgage insurance. Lenders commonly offer an array of loan packages requiring down payments of just 10 percent. But down payments can go down even further, as long as the insurance companies will cover it. These days certain buyers are eligible to put down as little as 3 percent.

Fannie Mae, in cooperation with loan insurers, has created the Community Homebuying Program, which requires a down payment of just 5 percent. Borrowers also have what's called a 3/2 option. The homebuyer must contribute a down payment of 3 percent and the remainder may come from outside sources, such as a gift from a friend, relative, government agency, employer, or charity. In some areas, certain employers are setting up programs to help their employees purchase homes. Qualifying ratios are also more liberal, typically allowing housing debt of 33 percent and total debt of 41 percent, perhaps more. To be eligible, you must make no more than 115 percent of the area's median income, though there are higher limits in the country's most expensive real estate markets. Freddie Mac has a similar program called Affordable Gold. These programs usually require that the buyers undergo some seminar-type training in home ownership.

The trend toward low-down-payment programs seems to be here to stay. GE Capital Mortgage Insurance companies has announced that it will make 97 percent LTV loans available to some buyers. These loans will carry a higher mortgage insurance rate than those in the Community Home Buyers Program.

Gifts

If you're lucky, a relative or friend may want to help you with the down payment. Some lenders restrict the use of gift funds for certain loans. Your benefactor will probably be required to sign a gift letter stating that the funds are a gift and come with no obligation for repayment. The lender may even want to do some checking on your donor to make sure the gift didn't really come from other sources. Lenders also place limits on the amount of the gift.

Depending on the size of the gift, your donor may have a tax liability. You should consult an accountant to be sure. If you absolutely must have the gift money for the deal to go through, make sure you're very clear about when the funds will be received. To truly protect yourself, you may want to write a contingency into the contract stating that the deal is off if the funds are not received by a certain date. Otherwise, if something happens, your donor dies or simply has a change of heart, you will be out the deposit. Getting a seller to accept this is another matter. In reality, you must simply be sure that the donor's offer is genuine and the money will come through for you. You should get the money into your account as soon as possible after your offer has been accepted.

Equity Sharing

Another way friends or relatives might help out is by taking an ownership interest in the house—what would be called an equity sharing or joint ownership agreement. You can do this with an outside investor as well. However, you might have a hard time structuring an agreement that allows a promising rate of return for the investor and still gives you a financial advantage over renting.

Equity sharing agreements can be structured in a number of ways and can be designed to favor the occupant or investor, as desired. Here are the basic principles: The investor brings cash and perhaps credit as well to the deal. The occupant lives in the property and also brings some capital to the deal. Typically, the investor will make the down payment or a significant percentage of it. The occupant then pays the closing costs. The occupant may make the full mortgage payment or the occupant might split the mortgage payment with the investor. The occupant also can pay rent to the equity investor. The agreement will outline key issues such as maintenance costs and the division of the proceeds of the sale. As a part owner of income-producing real estate, the investor is allowed to take a tax deduction if the property shows a loss, as it probably will. Certainly you can structure it that way. The investor will be able to write off depreciation of the property, a significant advantage. The occupying owner does not get this benefit. You cannot depreciate a property you live in.

As an example of the tax ramifications, let's look at a simple example. Note that this is in no way meant to be a model agreement. It simply illustrates some of the elements at play. Suppose the investor's, Sam's, down payment entitles him to a 50 percent ownership interest. The mortgage payment is $1,000. Julie, the occupant, will make three-quarters of the mortgage payment, or $750. She will also pay Sam rent of $300, which will cover his portion of the mortgage payment with $50 left over. Sam will be able to take an income tax deduction on 50 percent (his ownership interest) of the real estate taxes, as will Julie. They will both be able to deduct the interest payments for their shares of the mortgage. Sam will be able to depreciate 50 percent of the property. He also must count $3,600, Julie's annual rent, as income from the property. Still, he is certain to have an operating loss and therefore will have a deduction against his annual income.

Equity sharing can be win-win, allowing a relative to help you out and still receive significant benefits. There are dozens of ways to structure an agreement. As a buyer, you must ask yourself if it's truly better than renting, because you'll have to share some of the financial advantages of home ownership. You should consult a certified public accountant (CPA) if you're considering an equity sharing agreement.

Lease Option

In a lease-option agreement you lease a property with an option to buy it at an agreed-upon later date and price. Don't confuse lease options with lease purchases, where you are committed to make the purchase. You probably should avoid lease purchases unless you are absolutely certain you want to buy the property and will be financially capable of doing so.

Typically, you will pay a nonrefundable fee for the right of option. You also will pay an above-market rent. You and the owner can negotiate a portion of your rent that will apply toward the down payment if you exercise your option. Typically, that money also would be nonrefundable should you choose not to buy the property. Buyers should be wary of lease options. Typical sellers wouldn't consider one unless they're having problems selling, which indicates problems with the house.

Investors like lease options. They can attract quality tenants who will take care of the property, and history shows that chances are the sale will never go through. The reasons vary: The tenants may not be able to qualify for the loan; perhaps they haven't saved enough to cover the rest of the down payment; or maybe they simply lose interest. Meanwhile, during the term of the agreement, the investor is collecting above-market rent and gets to keep the deposit in the end.

That said, a lease option may still be a good strategy for you. You can lock in a price for the home, so you're protected against rapid appreciation in values. With a confirmed price and timetable, you can establish a savings plan that will make the purchase possible when the option comes up. Finally, you have time to try out the house and the neighborhood with no obligation to buy.

Seller Financing

If you're working in a buyer's market you may be able to get some financing help from the seller. When sales are sluggish, sellers are more anxious about making the sale than necessarily getting all their money out at once. On the other hand, if sellers are confident that buyers who do not need financing will come around, they won't be willing to deal.

The best scenario for seller financing is when you are a creditworthy borrower but lack the funds to cover your down payment and closing costs. The seller is taking a risk in lending you money. Arguably, it's a greater risk than for a lending institution, which hedges its bet by offering thousands of loans. The seller will be just as strict as any lender in qualifying you.

It is not uncommon for the seller to pay closing costs and points for the qualified buyer. In the grand scheme of things, these costs aren't a huge amount of money, not from the seller's standpoint. But for the buyer who's chasing every nickel to scrape up the down payment, it may be the difference in being able to afford the home.

☞ **Money-$aving Tip #39** *Here's a strategy for negotiating the purchase when you and the seller are close to an agreement, but you have maxxed out on your funds. Suggest to the sellers that you will go higher on the price if they will cover your points and closing costs. The sellers then aren't really losing money. In essence, you are financing closing costs by rolling them into a larger mortgage.*

Another form of seller financing is called the 75/10/15 loan, which is considered a conventional loan by the secondary market. The loan covers 75 percent of the purchase price, the buyer puts 10 percent down and borrows the remaining 15 percent from the seller. This is an example of using a second mortgage from the seller along with a conventional loan. The buyer avoids paying for private mortgage insurance because the secondary market considers this a conventional loan. The buyer also may be able to negotiate an attractive interest rate and terms from the seller, and will avoid the origination fees on the second mortgage. Your first mortgage lender will factor in payments on the second mortgage in determining your qualifying ratios.

The seller gets a good chunk of cash immediately. This makes for a good deal whenever cash flow is tight for a buyer, but funds will be available a few years down the road. Entrepreneurs with money tied up in a promising business, for example, may see this as a solution. Buyers also may use it to avoid the higher rates for a jumbo mortgage, which is an amount larger than the secondary market guidelines for a standard mortgage.

Another financing possibility is *assumption* of the seller's mortgage, which means basically taking over the obligations of the mortgage note upon purchase of the property. Most mortgages issued in the past several years have a due-on-sale clause, which means when the property is sold the mortgage must be paid off in full. However, FHA, VA, and some adjustable rate mortgages issued in the mid-1980s and earlier are assumable. The rules of assumption vary. Sometimes you will have to qualify with the lender, but not always. You should definitely seek legal advice before any loan assumption. You may be depending on the seller to make payments to the lender, or vice versa. You need a clear understanding of where risks lie in case of default. Most assumable loans are over ten years old. Because the home may have appreciated in that time, you will need extra cash. Traditional mortgage lenders will not issue a second mortgage on top of an assumed loan. If you need more cash, you'll be depending on financing from the seller.

You may run into a seller who owns the house free and clear and is willing to serve as your banker. An attorney's review is an absolute must for such an arrangement, which is called a *land sales contract* in some areas.

Seller financing is an option that can be win-win, but you must structure the deal carefully. You should have a third party hold payments in escrow. Qualification should be thorough, and payment schedules feasible. Balloon payments are a danger point. The agreement should specifically address remedies in case of default. Both sides should have legal representation.

How *Your Agent Can Help*

Some fortunate buyers have no trouble putting 20 percent down and qualifying for a mortgage. For many first-timers, however, a creative solution is in order. If you're cash-strapped, get together with the agent to brainstorm strategies.

- If you're in the market for a fixer-upper, timing is everything. The agent can get you into these properties as soon as they hit the market.
- If you're interested in Fannie Mae or REO properties, agents will help you search them out. They may know brokers in town who specialize in these properties.
- Agents also can assess the feasibility of pursuing seller financing in your market. As you look at properties, the agent can clue you in to sellers who may be receptive to an offer involving seller financing.

One of the most common cash-strapped strategies is a low-down-payment mortgage program. Agents can advise you on the rules of the FHA and VA loan programs. They also will know which lenders are the best to deal with if you're at the limits of qualification guidelines. Finally, they are your best source of information for special programs like those described above. Community groups, nonprofits, and city agencies are often involved in home affordability programs. For instance, to help revitalize blighted areas, some large cities are offering substantial subsidies for anyone willing to move in and fix up a property. Amazing opportunities exist, but you have to know where to look. These programs usually don't have the money to mount large publicity or marketing campaigns. It's up to you to find them, and your agent can help.

Commonly Asked Questions

Q. *Does a house have to have major structural problems to qualify as a fixer-upper?*

A. Not necessarily, but as a practical matter, probably. What makes a fixer-upper is in the eye of the beholder. The seller must agree that the property sorely needs work and be willing to make price concessions. A crack in the foundation can be pretty convincing here. But consider a home, for example, in one family for the past 40 years and owned by an elderly widow. The decorating is probably not to your taste. The kitchen and bathroom probably need to be modernized. The electrical system will need updating to handle your appliances. But from the widow's perspective, everything works. The roof doesn't leak. The toilet flushes. The basement is dry. She won't consider your estimate of remodeling costs to be a legitimate deduction from her asking price. Why should she pay for your tastes?

Q. *Why won't a lender allow me to repay a gift from my parents? If I ran into financial problems, they would just let me slide.*

A. The lender would see this as a second loan, an additional liability that could interfere with the payment on your mortgage. True, in most family loans the borrower will simply skip payments if times are tight. (A reason not to lend money to family members.) But the lender is not in the business of investigating the motivations of the parties involved. They simply want no other debt obligations on the house.

Q. *Why would a seller pay closing costs? Isn't that the buyer's responsibility?*

A. Yes, it is. But sellers also recognize that the down payment is the biggest obstacle for homebuyers, particularly first-timers. They won't pay points and closing costs out of a sense of charity. But if they think that's what they need to do to make a sale, they will.

CHAPTER 10

Shopping for a Mortgage

For many years choosing the right mortgage was a breeze. You had no choice—fixed rate, 30 years, take it or leave it. Even the adjustable-rate mortgage (ARM) is a relative newcomer, devised when banks were paying high interest rates on deposits while servicing mortgages originated years earlier at much lower rates. Banks introduced ARMs as a way to minimize their risk in a period of rising interest rates. Borrowers enjoy the same benefit of flexibility and can hope for their upside of declining rates.

Nowadays, choosing between adjustable and fixed rates is just one of many decisions when selecting a mortgage. Lenders are constantly coming up with new variations on loans. Shopping is a more difficult, yet rewarding, process. You'll be able to find a loan that best suits your financial situation. You may end up with the plain vanilla 30-year, fixed-rate loan in the end. But not until you've assessed the more exotic flavors. What you should remember is that you are shopping, and to carry that metaphor a step further, your goal is to "buy" a mortgage. Proportionally, the mortgage is a bigger purchase than the house, for which most buyers will put down cash of no more than 20 percent of the selling price. The rest of the money is from the mortgage. Choose wisely, and you can save thousands. From an investment standpoint, good financing is just as important as a good price in determining how much you will make on the property.

In Chapter 3, we looked at the mortgage application and underwriting process. The goal was for you to get a ballpark estimate of what you could afford, a logical starting point to your home search. In this chapter, we get down to brass tacks, considering your various options in mortgage types and how to compare both loans and lenders.

Special attention is given to the needs of first-time buyers. Two excellent programs for those struggling with down payment and qualification requirements are the loan programs of the Federal Housing Administration (FHA) and the Department of Veterans Affairs (VA). A common misconception is that FHA loans are for low-income people only. Though the size of the loans are limited to around the value of median prices in the area, there are no restrictions on borrower income. For a qualified veteran, the VA loan program is an excellent, and well-deserved, benefit. These programs are covered in capsule here. Both carry very specific requirements and guidelines that can change regularly, so interested readers are encouraged to consult lenders, agents, and other resources for detailed information.

Basic Mortgage Concepts

The homebuying process in general, and financing in particular, involve learning a language full of words you have probably heard before, but may not have fully understood. We encourage you to consult the glossary in the back of this book whenever you run across a word that puzzles you. Throughout this book we have used the word *mortgage* somewhat generically to refer to the financing of a real estate purchase. More precisely, a mortgage establishes a personal liability on the part of the borrower to repay a debt to the lender. The mortgage places a lien on the real property to serve as security for the debt. The mortgage is not the loan itself, but rather the document that secures it. If you default on the loan, the lender has the right to claim the mortgaged property under the laws of foreclosure. In some cases, a deed of trust (also called trust deed) secures the property. Whereas a mortgage involves a direct relationship between lender and borrower, a deed of trust involves a third party, who holds title to the property until the debt is paid. The foreclosure procedure is sometimes speedier in a deed of trust. We'll continue to use the term mortgage or mortgage loan to refer to the financing of real estate, though your loan may, in fact, be secured by a deed of trust. Either way, you better pay it off, or you risk losing the home.

Another key concept to mortgage loans is *amortization,* which refers to the gradual paying off of a loan through set payments over a given period of time. Each payment includes principal and interest. Amortization creates a schedule in which payments early in the life of the loan are almost entirely interest, with the share going to principal increasing slowly with each payment. The concept of amortization is important in selecting among some of the mortgage types discussed in this chapter. Certain loans carry the risk of *negative amortization,* which means that your payment is not enough to cover the interest due. The shortage is added to the principal, thus increasing the loan balance.

A third key concept is *discount points,* or simply *points.* A point equals 1 percent of the loan. The name comes from the fact that a lender will charge an up-front fee as a way to reduce, or discount, the interest rate during the life of the loan. This works well for buyers. For instance, if you know you'll stay in a house a long time and can afford to pay more points, you'll save in the long run from a lower interest rate during the life of the loan. However, this same feature can make it very difficult to compare rates. For this reason, truth-in-lending laws require that the lender disclose to you, among other information, the annual percentage rate (APR), which factors in the fees associated with getting the loan, including points, and expresses the true cost of the loan as an annual percentage.

Typically, the lender will set up an *escrow account* for payment on insurance and taxes. A portion of the borrower's monthly payment will be applied to the escrow account, from which the lender will make payments. The lender's security for the loan is the property. Unpaid taxes can create a lien on the property, which generally takes precedence over all other liens, including the mortgage. And a fire or other tragedy when insurance isn't paid up is an obvious problem for the lender. In total then, the mortgage payment will consist of principal, interest, taxes, and insurance, which is referred to by the acronym PITI. As discussed in Chapter 3, it is this figure that goes into the calculation of qualifying ratios.

☞ **Money-$aving Tip #40** *Lenders like to keep a cushion on escrow accounts. Federal regulations, which don't apply to all lenders, limit the cushion to two months of payments. FHA and VA loan programs permit only a one-month cushion.*

Tailor Your Mortgage

As you read the advice in this chapter, you'll see that it will be much easier to make decisions if you have a good idea of how long you'll stay in the property. Depending on your circumstances, it could be hard to be precise, but the issue deserves some serious thought. Are you a two-income couple who plan on dropping to a single income in a few years when you have children? Are your kids getting closer to college age? Would you like to start your own business in a few years? Or have you already bootstrapped a business that has tied up all your money but is beginning to produce big returns? In a world that's changing as fast as ours, it's hard to know where you'll be even in five years. But if you do have a plan, be sure to take it into account as you select a mortgage.

Looking for Lenders

Getting the right mortgage starts with putting time and energy into comparison shopping. The most common sources of mortgages are savings and loan associations (S&Ls) and banks. Other potential sources are credit unions, life insurance companies, mortgage bankers, and private investors. Some real estate brokerages have mortgage offices on-site, either their own or an outside company's. Of course, you are a ready-made market for their services, but you're not a captive market. If your broker performs this service, consider it just one of several sources to choose from. You should not feel compelled to use your broker's loan services just because the broker helped you buy the home. Compare his or her mortgage services on an equal footing with your other choices.

Mortgage brokers are intermediaries, matching borrowers and lenders. Real estate brokers may perform this service as well. They have the advantage of knowing more about the available loans than any buyer possibly could. Typically, the cost of a mortgage broker is covered by the lender, and you probably won't have any outright service charge. However, the lender will be charging you fees. You should scrutinize the lender's fees to make sure the cost of the broker isn't buried in some other excessive charge. No lender would undercut mortgage brokers with whom they do business, so you don't have any advantage going directly to the same source. In addition, brokers use sources that don't deal directly with consumers. Still, it is prudent to do some checking into the market yourself to be sure your broker is truly offering a good deal. Be sure your mortgage broker agrees not to be paid until the loan has been secured.

☞ **Money-$aving Tip #41** *Because of their extensive market knowledge, mortgage brokers can offer valuable assistance to buyers who may have difficulty qualifying with many lenders.*

The real estate pages of most local papers usually list the best mortgage rates available. Follow these rates for a month or so and make a list of lenders who are consistently competitive. Buyers with computers and modems can do some mortgage shopping online (see Chapter 15). Consider these sources to be a starting point. You may be able to get basic rate information from a database or Web site, but you really should be in direct contact with a lender to learn about its fees and application process. Otherwise, you may end up comparing apples and oranges. In addition, rates can change daily. Even Web sites do not update that frequently.

You should be collecting general information about the market and compiling your short list of sources while you're house hunting. When you're ready to actually apply for a loan, you'll want to canvass your lenders over a short period of time, preferably the same day.

When you call a lender, ask a lot of questions. The goal is to eliminate surprises and compare on a range of criteria. Don't choose a mortgage purely because of the lowest rate. There are other factors to consider—such as fees, servicing policies, and points to name a few. The APR (annual percentage rate) is a good standard of comparison because it factors in the fees.

Ask the lender about the following fees:

- The application fee and what it covers. It may, for instance, include the cost of the credit report.
- The credit report charge
- The cost of the appraisal
- Any and all other fees
- The annual percentage rate, an indicator of the real cost of the loan. The lender is not required by law to supply this information until after the application, but that's too late for comparison shopping. Always ask up front what the APR will be.

Your "other fees" question may need a follow-up question or two. Consumer advocates call these junk fees. In the lending industry they go by such names as document preparation charges, administrative fees, processing charge, or other labels. You may wonder why the lender needs to nickel-and-dime you with all sorts of charges when over the life of the loan they'll make hundreds of thousands of dollars in interest. Chances are that the lender will turn around and sell the loan on the secondary market. It therefore looks for some profit at the origination. You should question any fees you don't understand. You also should consider all charges to be negotiable. You may be able to put the closing costs into the amount of the loan, for instance. Of course, it might be easier to go to another lender who charges less, which in itself is a negotiating position.

Down Payments and Private Mortgage Insurance

The assumption so far in this book is that coming up with the down payment is going to be a challenge. This isn't the case for everyone. If you've won the lottery or are the beneficiary of a generous trust, you might be considering paying cash for the home. Many others are somewhere in between. For those with the luxury of a choice, how much money should you put down?

The traditional real estate loan requires a down payment of 20 percent of the purchase price. For smaller amounts you will have to buy private mortgage insurance, also referred to as PMI or simply MI. (FHA and VA loans are exceptions, but from the lender's standpoint they work similarly because they guarantee the loan.)

PMI insures the lender against default of the loan, but the borrower pays the premium. You will have to qualify for PMI. If you are able to qualify for the loan, it shouldn't be a problem. But there are no guarantees. The cost of PMI depends on such factors as the size and type of the loan, the amount of the down payment, and what the extent of coverage is. You will have to ask lenders what the PMI cost is for specific loans. The premium is almost always paid on an annual basis, though it is possible to make a one-time lump sum payment for a multiyear contract at the time of closing. If you pay on an annual basis you will usually make an up-front payment at the time of the close that is equivalent to 12–14 months of premium. The cost of the annual premium will then be averaged over your monthly payments.

This example shows the effect of PMI: The Millers buy a house for $130,000 with 10 percent down, or $13,000, and a loan of $117,000; the PMI premium is 0.36 percent and 0.49 percent must be paid at the close. The Millers would have an extra fee of $573 at the close ($130,000 × .0049). The annual charge for insurance would be $421 ($117,000 × .0036), or $35 a month. So if you need PMI, it's a great deal. For a relatively small amount of money, you are able to get into the house with a small down payment and still get the rates of a conventional mortgage. If the Millers stay in the house for seven years, the total cost of the insurance would be about

$3,500. Not small change. So if the Millers are able to make a down payment of 20 percent, or $26,000 in this example, the cost of PMI should enter into their decision.

We all have preferences for risk in our investment. The lower the down payment the higher your leverage and your risk, and if the market goes your way, the higher your return. You'll be able to take advantage of OPM—other people's money.

Leverage is a great thing when the market rises, but there is the downside risk to consider. Circumstances may force you to sell at a time when property appreciation has been slow or falling. At worst, you may get caught with a mortgage that's bigger than the value of your house at a time when you don't have the luxury of waiting the market out.

By investing the full 20 percent, you bear the opportunity cost—what you might have done with the extra money had you made a smaller down payment. You should ask yourself what rate of return you can expect from other investments. Not putting the full amount down costs you not only the PMI premium, but also the interest on the additional 10 percent you're borrowing. Whatever you don't put down goes into the mortgage amount. An offsetting factor is that the interest rate is tax deductible. If you're paying a rate of 9 percent and you're in the 28 percent tax bracket, your money needs to make at least 6.5 percent—[9% − (.28)9%]—in another investment for the smaller down payment to make sense. And, of course, this would ignore the PMI cost, so it should really be a bit more. You're left with more questions on alternate investments, such as will it be a bull or bear stock market?

The decision comes down to your attitudes about money, your financial self-assessment, and your plans for the future. Would you feel better making a small down payment and keeping cash in reserve for home projects or just for a rainy day? Or do you feel safer with the traditional 20 percent down, insulated from slumps in the real estate market? A larger down payment also will decrease your mortgage amount, and thus your monthly payment. Perhaps that cash flow advantage is important to you.

Choosing between 30-Year Loans and Shorter Terms

Many lenders are offering mortgages with shorter terms, such as 15 or 20 years. These loans offer substantial interest savings to borrowers. For example, a loan of $100,000 at 9 percent over a 30-year term would carry interest charges totaling $289,800; the same loan over a 15-year term would cost $182,700 in interest. In this case, you would save more than $107,000 with the 15-year loan.

As an added incentive, the interest rates are sometimes lower on 15-year loans. In actuality, you would therefore save even more in the above example. On the other hand, it will be more difficult to qualify for a 15-year loan because your payments will be higher. Your monthly principal and interest payment on a $100,000 mortgage at 9 percent over 15 years would be $1,015, as opposed to $805 for the 30-year loan. For a lot of families, that $200 will make all the difference in whether or not they qualify for the mortgage.

The larger monthly payment of a 15-year mortgage, attractive as the savings may be, could become a burden if you fall into tough times. There is a better option. First, confirm that the mortgage note does not allow *prepayment penalties,* which are charges for paying off the loan before its maturity. Many states now outlaw prepayment penalties, but be sure to ask about these charges when you canvass lenders. Prepayment penalties are not allowed in FHA and VA loans.

☞ **Money-$aving Tip #42** *Contact your lender before starting any prepayment program and check the rules on how extra payment should be submitted. Lenders may apply the money to your escrow account for insurance and taxes if procedures are not followed.*

If there are no prepayment penalties, you are free to tailor your own accelerated payment program. In the case of the above example, you could simply add $200 to every payment with instructions that it be applied to principal. Another method is to apply a small additional amount with each payment, perhaps as little as $10, during the first year of the mortgage, and increase the extra payment incrementally through the term of the mortgage. The advantage of the prepayment method to shortening the mortgage term is the peace of mind of knowing that if trouble strikes you can simply revert to more comfortable payments based on 30-year amortization.

Biweekly Mortgages

Some lenders are now offering biweekly mortgage programs. Instead of the traditional monthly payment, the borrower makes payments of half the monthly mortgage amount every two weeks. A mortgage amortized on the 30-year schedule will actually get paid off in 18 years. As an added bonus, you may be able to get a lower interest rate. Because your loan will be paid off sooner, the lender is taking less risk. Essentially you are making an extra mortgage payment every year. Instead of 12 monthly payments, you're making 26 half payments over the course of the year.

The biweekly program has a disadvantage for both the borrower and the lender. Simply, it's just more work to administer. The borrower has to be attentive to writing out that check twice a month, and the lender needs to do the work of processing it. There are a few ways around this. As part of the program, a bank may require that you maintain an account, which can be used for transferring money. You also may be able to set up an automatic payment program. Sometimes intermediaries coordinate the transfer of funds, in which case extra fees will be involved. You may very well have to pay more for the privilege of a biweekly mortgage. The lender does have a legitimate claim to increased administrative expenses. The lender will pitch these fees as a small price to pay for the big savings in interest.

No doubt the biweekly program is one way to save on interest without too much belt-tightening. It's not that much harder to put out half the money every two weeks than the full amount monthly. Nonetheless, it does affect your cash flow, which might be tight. And how much you're willing to pay for this benefit, if anything, is another question. Just as in the case of the length-of-loan term, prepayment allows you to achieve these same benefits on your own schedule. You may choose to make one extra payment a year and apply the full amount to principal. You could time the payment with an inflow of cash like an annual bonus or income tax refund. In a lean year, you can forgo the payment. When times are good, you might want to accelerate the loan further.

Fixed-Rate Mortgages

Fixed rate refers to an interest rate that will not change during the life of the loan. The appeal of the fixed-rate mortgage is its stability. It can be reassuring to know that your monthly payment of principal and interest will not rise for the life of the mortgage.

Fixed-rate mortgages usually carry a slightly higher interest rate than adjustable-rate mortgages, because lenders want protection against rate increases just as consumers do. The higher rate means a larger monthly payment, making it more difficult to qualify for a fixed-rate mortgage.

Adjustable-Rate Mortgages (ARMs)

The most basic advantage of adjustable-rate mortgages is that you benefit from declining interest rates when they are declining without having to go to the trouble and expense of refinancing, as you would with a fixed-rate mortgage. Another advantage is that the introductory rates are lower, making qualification easier. If interest rates are high in relation to historic patterns, ARMs are an attractive option. But be careful about getting into the business of predicting interest rates. Even leading investment analysts and economists don't always accurately predict where rates are headed. Just because interest rates are high does not mean that they cannot go higher. Nonetheless, in times of inflation the ARM will be more attractive. Even on the downside, if interest rates do continue to rise, it is quite possible that your income also will rise, though probably not as fast.

It can be misleading to compare ARMs to fixed-rate loans based strictly on interest rate. The initial interest rate on an ARM is often artificially low. This is known as a teaser rate. It is a fine feature, but dangerous if you are not prepared for a rate increase. Your mortgage will specify when the first adjustment will occur, which may be as soon as three months. For an accurate comparison of adjustable-rate mortgages, use the true rate, which as described below will be tied to a specific index. Following are descriptions of more key features to consider when shopping for an ARM.

Caps. A cap is the maximum amount the interest rate can rise on any one adjustment period or during the life of the loan. For example, a lender might quote a rate of 8 percent with a cap of 5 percent for the life of the loan and 1 percent per adjustment. You would know that the interest rate would never exceed 13 percent and would take at least five adjustment periods to reach that level. This feature adds a bit of predictability to ARMs and allows you to calculate the payments for worst-case scenarios. The combination of teaser rates and caps make the ARM an attractive option if you plan to stay in the house for a short period.

Adjustment periods. ARMs come with different adjustment periods. Most lenders prefer annual adjustment, but the periods do vary. What's best for the borrower is a low cap and a long adjustment period. Lenders are introducing a new variation of the ARM where the first adjustment is not made for a long term, such as five or seven years, and after which adjustments follow a traditional annual schedule.

Payment caps. In addition to capping the amount the interest rate can rise in one adjustment period, some mortgages cap the amount your payment can go up. For example, a loan with a 3/1 cap means that the interest rate could go up 3 percent each adjustment period, but the payment can only go up 1 percent. While this may seem to be a nice security, it also could lead to the dangerous situation of negative amortization. The amount of interest in excess of what is allowed by your payment cap will be added to the loan amount. So your loan will actually be getting bigger, not smaller, even though you continue to pay.

Indexes. Your interest rate will be tied to an index. Your bank will add a margin, usually two to five percentage points, to the index of measurement. Common indexes are Treasury securities, T-bills, the Federal Home Loan Bank Board (FHLBB) contract interest rate, and the 11th district cost of funds. You can research indexes in *The Wall Street Journal* or ask a lender. Try to come to an understanding of how rates fluctuate. Some, such as a six-month T-bill, are quite volatile and therefore best to use when interest rates are very high. Others, such as five-year Treasury securities, are very stable and would be best if interest rates are low. The five-year Treasury securities index is usually only slightly lower than the prevailing fixed rate, so taking into account the margin, the price for this index's stability might be a less competitive interest rate. Of course, remember that the index is not your rate; you must add the amount of the margin.

And new twists to come. It would be impossible to describe all the varieties of ARMs available, and new features regularly crop up. One variation is a convertibility option. After a period of time it allows you to convert to a fixed rate. This might be a good option if you'd be more comfortable with a fixed rate, but you need the easier qualification of an ARM. Once again, there is no replacement for shopping around to get a good idea of what's available.

FHA Loans

The two main advantages of FHA loans are low-down-payment requirements and lenient qualification ratios. The down payment amount can fluctuate, but has been about 3 to 5 percent. Qualification ratios are usually 29 percent for housing debt and 41 percent for total debt.

FHA loans can be fixed- or adjustable-rate and come in a lot of varieties. It is essentially an insurance program, and the borrower pays the premium. There is an up-front mortgage insurance premium (MIP) charge, which is fairly high (about 3 percent, but subject to change), but can be added to the loan amount. There also is an annual renewal premium, usually about a ½ to 1 percent. Critics argue that the FHA program is becoming too expensive, and that alternatives offer a better deal. The process also is more complex than it used to be.

FHA loans can be a good option for a lot of middle-income buyers who are moderately qualified but don't have much money for a down payment. The FHA sets maximum loan amounts that are based on property values in the area. You will be able to use these loans to buy houses priced at your area's median value or somewhat above. The property must pass some fairly strict inspections by the FHA, which is all for your protection.

VA Loans

For those veterans who are eligible (you must have been on active duty during specified periods), the Department of Veterans Affairs loan program is a great entitlement, helping thousands of veterans buy reasonably priced homes. There is little or no down payment necessary, loan fees are modest, and qualification criteria liberal. Interest rates are relatively low and the VA determines the maximum interest rate the lender may charge and sets no limits on the amount of the loan. Lenders, however, may have their own maximums. In effect, the maximum becomes the highest dollar amount of VA loan the secondary market is willing to purchase. The VA uses a debt-to-income ratio of 41 percent for qualifying. Both fixed- and adjustable-rate loans are available. The VA guarantees the loan, and the level of protection of borrowers' rights in a default situation will be found nowhere else. The VA loans also are assumable. Buyers must qualify, but they don't have to be veterans. That's the one way nonveterans can take advantage of the program.

As with the FHA loans, your property must pass an inspection. The application process also will take longer because of the bureaucracy. The first step is to obtain a Certificate of Eligibility from the Department of Veterans Affairs. Contact your closest VA office to find out if you qualify.

Other Mortgage Types

Below are some common variations of mortgages. Some may be of particular interest to first-time buyers.

Balloon Mortgages

A balloon mortgage is a good option if you are certain that you will be in the home for only a short period of time. The mortgage will carry a provision for a balloon payment covering the full balance due at a defined period of time. Typical length of the loan would be five or seven years. The monthly payments prior to the end of the loan are based on a 30-year amortization schedule. Because the lender is making a short-term loan, balloon mortgages generally have attractive interest rates. The downside is that if you are still in the home at the end of the term, you will have to come up with a substantial amount of money. If you don't have that kind of cash on hand, you'll have to refinance. There is always the risk that your financial situation would make qualification difficult or that you'll be forced to refinance during a period of high rates. If so, you may have to sell the house. If it looks like you'll still be in the house when the note comes due, don't wait until the last minute to refinance.

The Two-Step

You can think of the two-step as adjustment light. Like an adjustable-rate mortgage, the starting rate will be somewhat below market, maybe by as much as C\, percent. Instead of an annual adjustment, however, the loan adjusts only once, after a set period of five or seven years, depending on the program. After this point, it's like a fixed-rate mortgage. The two-step is tied to a ten-year Treasury index, but cannot increase more than 6 percent. True, the worst-case scenario of a 6 percent increase is hefty. Imagine going from a 9 percent to a 15 percent mortgage. Still, the two-step is a good solution for first-time buyers. Statistics show that seven years is a typical period of ownership for a first house. Even with the possibility of a big adjustment, the two-step is much less risky than a balloon note. At least you are assured of a 30-year loan commitment.

The Buydown Mortgage

The buydown mortgage is a financing technique that lowers the monthly payment during the early years of the mortgage. The buyer, seller, builder, or other party pays a premium in the form of points that "buy down" or lower the interest rate for a set period of time. A permanent buydown, which is less common, reduces the interest rate for the life of the loan. A buydown is a common financing incentive used by new home builders. It also may be a way to use a relative's gift.

One example of a buydown is the 3-2-1 loan. Here the interest rate is 3 percent lower than the agreed-upon interest rate of the loan for the first year, 2 percent lower the second year, and 1 percent lower during the third year of the loan. After the third year the loan goes to the interest rate set in the mortgage note, and stays that way for the life of the loan. It's likely that this interest rate will be higher than other conventional loans in the market.

How Your Agent Can Help

Agents have always assisted buyers with their financing needs. Agents can tell you about loan programs that fit your financial needs. They may recommend lenders who are more forgiving of a blemish on your credit record or have programs with liberal qualifying ratios. Some agents will go so far as to help you fill out the application. Certainly they will give you speedy advice when you have questions about the information you need.

Agents also will follow up with the lender to make sure the underwriting process is moving along. They will make sure the appraiser gets in to see the property on a timely basis. They will check that documents necessary for the close are being prepared. It is in their self-interest that the sale go through without a hitch.

Don't take these services for granted. Agents may not volunteer them, but most will be more than willing to provide whatever help you need.

Commonly Asked Questions

Q. How do I know my lender isn't keeping too much of my money in the escrow account?

A. You are right to be concerned about escrow account balances. The lender earns interest on these funds. There have been abuses. You should check your monthly statements to make sure the escrow portion is properly credited. At the end of the year tally the expenditures and the balance kept in the account. If the cushion is larger than two months, or one month for FHA and VA loans, you may have a problem. You should also be certain that escrow money is properly credited after a sale or refinancing.

Q. Can I choose my own PMI company?

A. PMI is an odd insurance product. Though the borrower pays the premium, the lender is really the client. Your lender probably will have one or two companies that it works with and will determine where you apply. If you are turned down, the lender might forward your application to other companies, but not necessarily.

Q. What if I'm unable to pay my mortgage at some point during the loan?

A. First, you need to make every possible attempt to pay the mortgage loan. Work with all other creditors before you try concessions with the lender. Commonsense advice is to call a loan officer as soon as you suspect that you may be unable to pay. In the best-case scenario, you may be able to work out a schedule of smaller payments to be made up at a later date. This is called forbearance, and the lender is basically passing on its right to take legal action. The mortgage note obligates you to make the monthly payment in full. Anything less is default. Consider yourself lucky to work out such an arrangement and naively optimistic to expect one.

Lending institutions tend to be large bureaucracies. The unfortunate reality is that you may not be able to get the attention of a bank or S&L until after you've already defaulted.

Perhaps even for a couple months. Nonetheless, it's good advice to try to establish contact with the lender as soon as possible in the midst of financial trouble.

Taking on a mortgage loan obligation is what makes most first-time buyers nervous. It is indeed a responsibility to take seriously. However, you can take heart that even in the worst of times foreclosure rates are extremely low. First, foreclosure is always the last resort for the lender. Second, if a lender considers you a good risk, you probably are. Lenders are, after all, notoriously conservative.

CHAPTER 11

Your Home Inspection

You've just moved into a house that, if not everything you've dreamed, is at least just right. The first thing you do is paint the kitchen. You don't understand how the previous owner could live with that color. You even put off unpacking boxes, except for what you absolutely need. When it's done you love it. You love the whole house. You prepare your first home-cooked meal in a week and move on to the rest of the chores of settling in. A few days later you're doing your first load of laundry and notice a pool of water near the downspout drain. Then you see water damage at the ceiling. You check the walls on the first floor and second floor. You call in a plumber to find the problem. "There's a crack in the waste pipe from the second-floor bathroom," the plumber says, after making small holes in the wall below the kitchen and bathroom sinks. "I'll probably have to knock a hole in the kitchen wall to get a new one in." Not the kitchen wall, you think. Your muscles still ache from the painting. Why didn't we see that, or the inspector either, you wonder. "They probably didn't use that sink," the plumber says.

The story is hardly unusual. Sometimes the responsibilities of maintaining the home will hit you like a ton of bricks soon after you move in. You probably will get some surprises. The trick is to avoid the big, expensive ones. The old adage *caveat emptor,* let the buyer beware, still holds, though things are getting better. Most states have *seller disclosure* laws, which require the seller to fill out a standard form or otherwise reveal in writing all the known defects in the property. Both you and the seller will sign the document. But don't let seller disclosure rules lull you into a false sense of security. All localities also have laws against breaking and entering, yet the home alarm business is booming.

A standard contingency clause in purchase contracts today is that the house must pass a professional inspection. Of course, it's the buyer's responsibility to arrange and pay for the inspection. It will cost you a few hundred dollars. Because buyers have the protection of the contingency clause, most inspections take place after you and the sellers have agreed on a price and signed a purchase agreement. That way you won't waste money until you're sure you and the sellers can come to an agreement.

The purpose of the inspection is to identify major problems as well as minor repairs you will need to take care of. Some minor work is to be expected when you buy a home. You may need to do a bit of caulking and painting or clean the gutters. Replacing the roof or buying a new furnace is another matter. If the inspector uncovers major problems, you will reopen negotiations with the sellers. You may require that the sellers make repairs or reduce their price to compensate for your repair expenses. You also have the right to back out of the agreement.

You'll want to do your own, less formal, inspection of houses as you shop around. It's better if you can identify problems on your own, before you make an offer. Typically, you'll look at houses for weeks, or maybe several months, before you're ready to put an offer in. When you see a home you like, you'll return, possibly more than once. In this chapter, we'll give you some ideas on what to look for in these subsequent, eagle-eye visits. We'll also discuss the professional home inspection process. Under no circumstances should you forgo a professional inspection. The only exception might be if you're a construction expert yourself. Even then, a second opinion is worth considering before making such a substantial investment.

The Once-Over

In previous chapters, we discussed the evaluation of house designs and neighborhoods. You'll probably go through many houses, imagining yourself living in each, judging its features against your likes and dislikes. You'll find one or two where you think, this might be it. Then you'll wonder, is there anything wrong with the house? We'll call that the once-over, though twice- or thrice-over might be more appropriate. On your return visits, you'll scrutinize the property thoroughly, looking for reasons to not buy it or to knock down the owner's asking price.

Structural Problems

You may be able to identify structural problems from the exterior of the house. A pair of binoculars can be a big help in your inspection. The roof should not dip or sag. That may mean it is overloaded. Check that there are no large gaps between the house and an attached garage, addition, or porch. Make sure masonry chimneys don't lean. Balconies can be quite susceptible to wood rotting. You also should look closely at porches.

The lines of the house should be straight. Look for any bulging or leaning at the sides. Stair-step cracks on the inside or outside of the foundation indicates that the house has settled unevenly. You also may see these cracks in brick exteriors.

Inside the house you should check that doors and windows move properly in their frames. If not, there may be some movement in interior walls. Put a marble on each of the floors. If the marble moves on its own, there may be problems with the floor itself or the foundation it rests on.

Drainage Problems

Water wreaks havoc on house building materials. Look for signs of water damage around the house. Gutters perform the key function of directing water away from the house. Check that the ground around the house is graded so that water moves away. After construction, soil is piled around a house. If it settles, you might have the opposite effect—a moat around the house. Your home is your castle, but a moat you don't need.

Inspect the basement walls thoroughly; that is where a water problem on the outside will show itself. Brick walls will show a white substance called effervescence. Look for other signs of water penetration, such as mold, mildew, or water stains. Be suspicious of a fresh coat of paint; it may be there to hide a problem. Moisture in a basement or crawl space will evaporate, working its way through the house and possibly causing problems in the roof or attic.

Mechanical Systems

Pay close attention to the house's heating and cooling, plumbing, and electrical systems. The older the house, the more prone to problems these systems are, whether due to wear or just being out of date with modern needs.

You should be aware of what kind of heating system the house uses. Check the age and condition of the furnace or boiler. Ask owners about annual maintenance procedures and what work they've had done to the system. Look for obvious design problems, like vents on the ceiling rather than the floor. Ask for the owner's average monthly utility bills.

If the water comes from a well, check its location and confirm that the water pump is in good working order. The way to test water pressure is to turn on the taps and then flush the toilet. This procedure is especially important in two-story homes. Check for signs of leaks in supply and waste lines.

The standard for electricity into a house is 100–200 amps. Check the condition of wires coming into the house. Pay attention to the number of receptacles in rooms and their location. Adding more receptacles is relatively inexpensive, but a major upgrade to the electrical system could easily run into a few thousand dollars.

Inside the House

Check that doors and windows are adequately weatherstripped to keep out the elements. Look for signs of moisture damage around windows. This is an indication that too much cold air is getting to the inside surface of the window, causing condensation.

Look for brown spots, peeling paint, or other signs of water damage on ceilings. Look at the condition of the walls. Structural problems also will show on the inside. Some cracking of plaster is normal, but drywall should not crack. Floors should be reasonably level, without sagging.

Look carefully in the bathroom and kitchen. There should be a ventilation fan in the bathroom to keep indoor humidity down. Kitchen ventilation fans also are important for removing odors and moisture. Make sure a range hood fan vents to the outside. Receptacles should be GFCI (ground fault circuit interrupter) wherever water is nearby. Look for signs of leaking pipes. Feel the floor for soft spots around shower stalls, sinks, and toilets. Kitchen appliances may or may not be part of your purchase. Take note of the brands and make sure they're in working order.

Check the attic for insulation and good ventilation. Bad insulation lets in the cold and also could cause moist attic air to condense, resulting in water damage. If the attic is not properly vented, moisture cannot leave the house. Ideally, there should be vents both high and low on the roof. It is difficult to remove the moisture from the attic. Exhaust fans in the house sometimes vent into the attic, causing excess moisture. Look for rusty nails or truss plate, mildew, rotted wood, and other signs of moisture damage.

☞ **Money-$aving Tip #43** *Homeowners warranty programs that insure new owners against property defects have long been used in the new home industry and are becoming more common in resales. These tend to protect the seller and broker from disgruntled new owners as much as anything. In fact, sellers may be lax about disclosure requirements when they have a warranty on the home. If problems crop up, you'll have deductibles and exceptions to deal with. There's nothing wrong with such a policy, but it is still just as important to hire a professional inspector.*

The Home Inspection Business

The business of inspecting homes is a relatively new one. Nowadays a home inspection is standard procedure for home-buyers. Not so many years ago a fraction of homebuyers would hire an inspector. All these new customers in an industry that lacks regulations has created a carpetbagger mentality, attracting many unqualified inspectors who are looking for easy money. Therefore, a recommendation from a trusted source who has used the inspector should carry a lot of weight.

☞ **Money-$aving Tip #44** *Avoid inspectors who also make repairs. They have a built-in bias and may be looking for problems that don't exist. Another example of a biased inspector is one anxious to refer you to relatives or friends, or anyone willing to give a referral fee. You should take note if unsolicited referrals are coming your way during the inspection.*

As discussed in Chapter 4, it's sensible to put membership in the American Society of Home Inspectors (ASHI) on your list of criteria. You also should check that the inspector carries insurance. Inspectors should be covered for any damage that may occur during the inspection. They also should carry errors and omission insurance, which protects against negligence in the inspection.

What the Inspector Should Do

No inspection can be as thorough and detailed as to catch everything that can go wrong with the house. For your fee of $250–$400, however, you should get the close attention of a knowledgeable home construction expert for a couple of hours. You and the inspector should agree up front on what the scope of the inspection is. At the minimum, the inspection should cover the major structural components and mechanical parts of the house—the foundation, roof, electric, plumbing, and heating and air-conditioning systems. Most inspectors also will check the condition of insulation, doors, windows, gutters, and appliances.

FirePlace

The inspector should come with some basic equipment: a flashlight, screwdrivers, circuit tester, and a ladder. Others also might use binoculars, a tape measure, and a builder's level. You would expect old clothes, too. The inspector should be willing to get a good close look at crawl spaces or attics.

Finally, you should get a detailed written report with specific comments on the various systems and components of the house.

Make the Most of the Inspection

You should absolutely be present at the inspection. How else can you be assured that the house has undergone a detailed examination? It's also important for you to be fully knowledgeable of the house's condition and working systems. Take notes. Don't rely completely on the inspector's report. The inspector may make many useful comments that won't find their way to a checklist-style report. Some homebuyers tape the inspection, and it's a good idea. If there's anything you're unclear about later, you'll have a complete record to refer to. When you go to the owners to renegotiate an offer, you'll be able to accurately describe the problem.

Don't be shy about asking questions during the inspection. Of course, you don't want to distract the inspector from the business at hand. If the inspection becomes a short course in home construction, the quality may suffer. But you should understand everything the inspector says about the house. Ask questions until you do.

Problems uncovered in the inspection can open the doors to renegotiation. You do want to be reasonable. The owners shouldn't have to repaint a room because a bit of paint is peeling. On the other hand, if the peeling is due to a leaky roof, you are entitled to either a price adjustment or the necessary repair work. If your contract has a contingency clause for home inspection, as it should, and the inspector has discovered problems that have changed your mind about the house, you have every right to back out. Inspections are meant to give you some protection from problems you overlooked at the time of the offer.

If you do end up buying the home, keep a copy of your report on file. It can become the starting point for your maintenance schedule. The inspector will probably find potential trouble spots you should keep your eye on. You should take a look at the report every so often as a reminder.

Environmental Problems

Potential environmental hazards in and around the home deserve special attention. Certain building materials that we now consider to be potentially dangerous at one time were widely used in residential construction. Some potentially hazardous materials are very easy to spot, such as asbestos insulation around exposed pipes. Others you'll have to test for, such as the odorless and invisible radioactive gas radon. If you're looking for a fixer-upper to rehab, you should be particularly attentive to potential risks. Whether completely gutting the building or simply scraping paint, you may be releasing dangerous dust particles into the air you and your family breathe.

Seller disclosure laws cover environmental problems. In addition, if you ask sellers about environmental problems in or around the house, they have a legal obligation to answer your question honestly. Because lenders share in the risk of a property, some are beginning to require environmental tests, such as radon screening, as part of the property evaluation process. Of course, a seller may not know about an environmental problem or may choose not to disclose it despite the law.

When you're selecting a home inspector, you should ask if he or she will help you identify common environmental hazards and if he or she has any special training or certification to do so. The federal government and other agencies publish plenty of free or low-cost information about environmental hazards. Your regional Environmental Protection Agency (EPA) or state health department is a good place to start. If you're having trouble locating these sources, contact your local library. If you suspect a serious problem, you may have to hire a specialist.

Below is information about three of the most common risks—radon, lead, and asbestos. You also should look for other potential problems around the home. You should be leery about properties close to gas stations. Leaking underground storage tanks can be a serious problem, but there's no way to really check on them. Another potential concern is power lines nearby. Some people believe that their electromagnetic fields could lead to health problems. The extent of the risk is still unclear, but the perception alone could cause problems at resale.

Radon

No home, new or used, in any part of the country is free from the risk of radon. According to the EPA, 1 out of 15 homes has a high radon level.

Ask the seller if the house has been tested for radon. If so, find out where and when it was taken, what the results were, and if any action was taken to correct high levels. You can use the results of the sellers' test or request that another test be taken. You should test in the basement, where radon levels are highest. If the house doesn't have a basement, use the lowest living area of the house. You can buy short-term testing kits in a hardware store. Make sure any kit you purchase indicates that it meets EPA requirements, and follow the instructions on the packaging. You also can consult a professional, who in addition to identifying a problem can advise you on the cost of correcting it.

Lead

Though the biggest lead problems are in low-income public housing, middle-class homeowners are in no way exempt. Before World War II, lead was common in house paints. About one-third of the houses built between 1940 and 1960 contain lead paint. It was not until 1978 that lead beyond miniscule amounts was banned in the manufacture of paint for residential use. So unless a house was built after 1980, there is a good possibility it contains lead paint.

Lead is harmful to many human body systems. Children, with small bodies and developing nervous systems and brains, are particularly vulnerable. With habits like crawling on floors and carpeting, playing near window sills, and putting things in their mouths, children also are more likely to come in contact with lead. Pregnant women should be just as cautious as parents of small children because lead is carried through the blood and can harm an unborn child.

Many people buy older homes, planning major remodeling or renovation projects. But renovating a home with lead-based paint can be dangerous without proper, sometimes expensive, precautions. Scraping, sanding, or heating lead-based paint will release harmful amounts of lead into the air. A fine dust will settle on floors or carpeting and will go right back into the air with vacuuming or sweeping.

Lead can be a difficult problem to correct. Painting over a lead-based paint is not sufficient. Removal should be done by qualified professionals. Another solution is to encapsulate the affected surfaces with a solid fireproof barrier, such as drywall. For windows and doors, replacement can be a cost-effective solution. Windows with lead-based paint can be particularly hazardous because the opening and closing loosens chips and dust particles.

If you buy a home built before 1978, the law requires that you be given ten days in which to obtain a lead hazard risk assessment or inspection, and your sales contract should contain a lead warning statement. You should be able to get a lead paint inspection for a few hundred dollars.

Another potential source of lead is the water, which might pick it up from the municipal water supply and service pipes or piping in the house itself. You can call your local health department for information about testing the water, which is sometimes offered as a free service.

Asbestos

Asbestos is a fibrous mineral and a frequently used material in housing because it is strong, durable, fire retardant, and an excellent insulator. Studies have shown that workers exposed to a large amount of asbestos showed an increased risk of cancer. It is less certain how dangerous the levels found in most homes can be. However, experts have been unable to prove that it is completely safe.

Since 1979, the use of asbestos in home construction materials has been reduced substantially. Prior to 1973, asbestos was commonly used in such items as floor coverings, ceiling tile, duct wrapping on heating and air-conditioning systems, insulation on hot water pipes and boilers, and fireproofing. If you're looking at an older home, chances are it does contain asbestos. The only way to be completely sure is to hire a qualified professional to survey the home and test the material. You also may be able to collect a sample yourself and send it to a lab for testing. Health departments or EPA offices can supply a list of labs and testers.

Asbestos is only dangerous when it is friable, that is, when deterioration causes it to loosen, crumble, and flake. Its miniscule fibers will then be released in the air and will lodge in the lungs when inhaled. If the asbestos is in good shape, there is little risk, and the best thing to do is to leave it alone. You can make minor repairs on pipe or duct insulation yourself, using paint, duct tape, or sealant. Proceed with caution. Always avoid scraping, sanding, sawing, cutting, or drilling any material that contains asbestos.

How Your Agent Can Help

Most real estate agents are not construction experts. However, they do look at a lot of houses. They also hear a lot of horror stories about house problems. After a while, they get an eye for spotting problems. Though an agent's insights are by no means a substitute for professional inspection, they may be able to help you spot potential trouble spots. Ask them questions about anything that arouses your concern. You should know that regardless of who the agents represent, they are required to disclose any material defects in the property—both those known or what would be considered discoverable by reasonable inspection.

Agents also can recommend inspectors. You should recognize a possible conflict of interest here. A home inspection that uncovers problems can kill a real estate transaction. When inspectors are getting a lot of referral business from an agent, they may develop a sense of loyalty to the agent. There should be no question that the inspector works for you. The inspector recommended byan agent may turn out to be excellent. What's important is that you screen these candidates as you would any other.

Buyer's agents can advise you on using the inspection report to negotiate concessions. Of course, they also will be your advocate in those negotiations.

Commonly Asked Questions

Q. Should I hire an inspector to look at a new home?

A. Yes. It is typical that before you close on a new home you will have a final walk-through to identify problems. The builder should have its own system in place for checking the quality of its contractors' work. But all too often, if problems occur after the deal is closed, it's hard to get the builder to make repairs. Your best protection is to hire an inspector.

Q. How much time will I have to get an inspection done?

A. That depends on how you write the contingency in the contract. But it will be a matter of days. Five days after the date of acceptance is common. That's why it's best to select your inspector before you actually need one.

Q. What are some of the most common problems with old houses?

A. Kitchen and baths often are in dire need of modernizing. These are a couple of the most important rooms in the house. Remodeling will give you a high return on the dollar but is still expensive. Plumbing can be clogged up. The iron and steel pipes that were common 50 years ago can jam up with rust and corrosion. Older houses are likely to have inadequate electrical systems. The wiring itself can last a long time if it has been properly fused, but you may need to increase service into the house or add receptacles or fixtures. Consider the typical number of electric appliances and devices in our homes today compared to those of 50 years ago.

CHAPTER 12

Negotiating the Purchase

It's the moment of truth. You're ready to make an offer. As a first-timer, you're probably not sure how you go about it. You know that this is different than a trip to your corner grocer. You aren't planning to knock on the seller's door and say, "I'll take it. Where do I sign?"

If you're nervous, that's normal. And it's good. Making an offer could land you in a binding agreement dealing with sums of money well beyond what you're used to. A little edginess will keep you alert. No need to go overboard, though. With your knowledge and experts at your side, you can deal with confidence. At this stage, you'll be working with the agent and your lawyer.

What Happens

As you've moved through the various phases of the homebuying process, you have laid the groundwork for making an offer. You know how much you can afford to pay. You've looked at various financing options and have an idea of the mortgage you want. You may have been looking at homes for a few days or a few months. Now you've found one that you want.

The process begins when you sign a written *offer to purchase.* This agreement outlines a number of key legal issues and negotiating points, which will be discussed later. You should take it very seriously. The seller might just say yes, in which case you have a binding agreement. You are not just running a price check to see what the sellers might take. An offer to purchase means just that.

Your agent will present the offer to the sellers or to the listing broker, who will in turn bring it to the sellers. The sellers have three options: they may accept the offer, reject it outright, or reject it with a counteroffer. What they do depends on how attractive your offer is. If you offer the asking price or very close to it without any major contingencies, chances are good they'll say yes. A little back and forth is common. If you're in the ballpark with your offer, the sellers are likely to counter.

As soon as an offer is countered, it's off the table. The ball is now in your court, and it's the counteroffer that is in play. You are released from any and all obligations of your offer to purchase, which started the negotiations. You can walk away. If you refuse the counteroffer, there is no deal. You don't revert to your initial offer. If that's what you want to do, then counter with your initial offer, which sends the owners the message that they've already seen the best you're going to do. It also tells them that you're inflexible, and may very well end the negotiations. When agreement between the parties is reached, each signs the offer and you have a contract of sale or purchase agreement.

Preparing the Offer

Because the offer is binding, it's important that your lawyer be involved. The safest course of action is to work with your lawyer from the start in writing the offer. In most states, real estate agents also can help you draw up the agreement. They will ask you questions and fill in the blanks according to your directions. You can take the offer to your lawyer for review before it is submitted to the seller. You also can deliver the offer but make it subject to your attorney's approval, which is your next best option. This common contingency allows the buyer to act quickly in a competitive situation while still providing the protection of legal counsel. We strongly suggest that your offer receives some form of legal review before the contract becomes binding. There is simply too much on the line and too much fine print for the first-time buyer to go it alone. In some places, like New York, an attorney must prepare the actual contract after the real estate agent has negotiated a "binder."

☞ **Money-$aving Tip #45** *While a "subject to attorney approval" clause affords some protection, it also backs your lawyer into a corner. If there is a problem, your lawyer's the bad guy, the sellers are on the defensive, and the agent will grumble that lawyers are deal breakers. The lawyer may be reluctant to work the finer points. You'll get the most out of counsel if your lawyer can review the offer before it is presented to the seller. Unfortunately, there is not always time. The danger is that somebody else may bid while you're awaiting your lawyer's review.*

Purchase agreements are available at the local Board of REALTORS®. In most places, no one single agreement is standard. Most likely, several varieties of forms will be available through the board and elsewhere. It's not sufficient to concentrate only on the blanks of the forms. You may consider most of the form boilerplate, but you should nonetheless read it word for word and be sure you understand everything. The purchase offer may come out of the broker's word processor. Boilerplate doesn't have much meaning when the forms are so easily changed in word processors. Just because something is boilerplate doesn't mean you can't make changes. Everything is subject to negotiation.

What Is in the Agreement

For starters, the agreement lays out who you are, who the seller is, the address and legal description of the property, and what purchase price is being offered. Those are the all-important basics. You'll lay out other details of the transaction, including terms that are up for negotiation. You'll also put in

some contingencies, which will be discussed separately in the
next section. Here are some important points your purchase
offer will address:

- *Earnest money.* This up-front deposit gets its name from
  its function in indicating to the seller that you are indeed
  serious. There is no hard and fast rule as to the amount.
  Common practice around the country is to put down 1 to
  3 percent of the purchase price. How much you put
  down depends on your negotiating strategy. It need only
  be enough to get the seller to deal with you. On the other
  hand, one strategy would be to accompany a low offer
  with a fat earnest money check, which gets the seller's
  attention and says "let's talk." Earnest money should go
  into an escrow account, not to the listing broker and especially not directly to the seller. Some localities have
  very specific regulations on how these funds are ex-
  changed and handled. Should the deal fall through, you
  want to be able to get this money back.
- *Timing.* Define how long your offer stands. You want to
  make this period short, a day or two, to avoid the situation where the owner sits on the offer waiting for a better
  one to come along.
- *Down payment.* You should indicate how much of the
  funds will come in cash. You can put in details of the
  financing as a contingency.
- *Closing costs.* Though most of the costs of the close are
  traditionally the responsibility of the buyer, you may
  want to bring these up for negotiation, as discussed in
  earlier chapters. Many localities charge taxes for the
  transfer of real property. You may want to negotiate that
  these are split.
- *What's included.* Typically, fixtures in the home stay and
  personal property goes. However, this area is frequently
  the source of disputes in transactions. You should specifically address major appliances—refrigerators, ranges,
  washers and dryers. Light fixtures, floor coverings, and
  window treatments often come up in disagreements. If an
  unusual lighting fixture is important to you, for instance,
  request that it stays. The seller may be planning to take it
  along, replacing it with a standard variety.

- *Taxes.* Real estate taxes for the current year are up for negotiation. A typical solution is to prorate them depending on occupancy.
- *Title and title insurance.* You will indicate how you will take title to the property. Married couples often choose *joint tenancy* (where available), which grants rights of survivorship. If one spouse dies, the other owns the property in full. Other possibilities are *tenancy by entirety, community property, tenancy in common,* or *individual tenancy.* How you take title has significant tax and credit ramifications. Unless you're experienced in tax and law matters, you should seek legal advice on this issue. Title insurance assures you that the title is good and unencumbered. It's frequently a seller's expense, though you may want extended coverage, which might need to come out of your pocket. You also will pay for mortgage title insurance, which covers your lender. See Chapter 13 for more coverage of title insurance.
- *Closing date.* Your agent can help with this information. You want to ensure adequate time to secure financing and for other details for settlement to be handled.
- *Occupancy.* You should define when the property becomes yours. If the sellers cannot move by that time, a rent should be established. If you're worried about the sellers staying too long, you can escalate the rent to provide an incentive. This is a bad idea in general because you become subject to landlord-tenant law. At worst, you may be forced to evict them. It's best to make a clean break—when you buy, they're out.
- *Walk-through.* It's recommended that you have the right to a final walk-through of the property before the close, to ensure that the condition has not changed significantly since the time of your offer.
- *Default and remedies.* The agreement is full of rights and responsibilities of both parties. The best agreements clearly address what happens if one side doesn't come through. Key here is the earnest money deposit. When is it returned and when does it belong to the seller?

Common Contingencies

A contingency is a provision requiring that a certain condition be satisfied or an event occur in order for the contract to become binding. In the previous chapter, we discussed the importance of the house inspection, which has become a standard contingency in real estate transactions. We also addressed the importance of a provision ensuring that your lawyer sees the offer before it goes into effect.

Contingencies must be carefully written because they frequently become a cause for dispute. The rights and obligations of both parties should be clear, and the consequences of conditions not being met despite good-faith efforts should be understood.

A common contingency is financing. A major factor in the cost of buying the house is the financing you can get. You know how large a loan you can qualify for at today's interest rates. But rates can change daily, sometimes quite dramatically. The financing contingency protects the buyer. You can specify the amount of the loan and the terms you require in order for the deal to go through. You may want to specify a type of loan—for example, you can only buy the home if you get FHA financing or a community homebuyers loan. There are no rules as to how you write the contingency. However, for your offer to be accepted, the sellers must feel confident that you can meet the conditions.

Another common contingency that you will not have to use still bears mentioning, the condition that you sell another property. Although you don't have the funds from a sale to work with when you're shopping for the first time, you also don't have to juggle buying a home with selling your current home as so many others do. Sellers are reluctant to accept an offer with this contingency, understandably. They don't want their deal to hinge on somebody else's. If you're in a competitive situation and are fortunate enough to know the other party must use this contingency, don't feel you have to compete on price.

The only limit on contingencies is your imagination. But the fewer the better. Expect any contingency you list to be accompanied by a deadline for removing it. If you are depending on a gift from a third party, you may have to write that into the offer. Perhaps Uncle John said he'll give you $5,000 for a house but wants to see it first. Try to bring him with you on a visit before the offer. Sellers will not like a contingency that's a big unknown. ("Who is this Uncle John guy anyway?")

About Negotiating

In some cultures, every trip to the marketplace involves multiple negotiations. In the United States, we rarely negotiate over prices. Exceptions are antique stores, garage sales, flea markets, and what so many people dread, the automobile purchase. Though negotiating over money is not all that common, we nonetheless negotiate our way through life. At its simplest, negotiation is communicating with others to obtain what you desire. You negotiate every day with your boss, coworkers, spouse, and children. In a typical negotiation both sides take a position. The sellers have their asking price. You have your offer. The further apart you are, the less likely an agreement will be reached. Ideally, after negotiations both parties will feel they've won. Lopsided agreements are more likely to fall apart.

Negotiations work best when egos do not get involved. Focus on the deal, and try to keep emotions out of it. That's the ideal. Unfortunately, selling a home is different than selling a lawn mower. You'll do best to acknowledge that it will be difficult to keep emotions out altogether, so you better account for them. Sellers have a lot of memories tied up in a house. They may associate some of the most important events of their life with the house. It's quite possible that if they don't like you, they won't like your offer either. The buyer also is not immune from irrational behavior. You may fall in love with the house to the point where you believe that none other will do. Hardly a good negotiating stance.

Remember that homeowners take great pride in their homes. You should avoid saying anything negative about the house in the presence of the sellers. Even comments that seem harmless, like redecorating you'd like to do, could cause the owners to take offense. Why would anyone want to touch such a beautiful decor, they'll wonder. You'll want to have your poker face on when you look around. If you're falling in love with the house, try not to show it to the sellers or their agent.

How Much Should You Offer?

The first step in negotiations is to gather as much information as possible. You've already given the house and grounds the once-over. You have a good feel for the neighborhood. You have a good idea of what sellers are asking for homes and some idea of what homes are actually selling for. Ask your real estate agent to pull some information on "comparables," or recent sales of similar houses in the immediate vicinity, from the MLS database. These listings don't tell all. For instance, you won't know the extent of repair work or remodeling that will be necessary in the home. Or, for that matter, anything about the condition. But the information will give you a frame of reference. Ask your agent to prepare a formal Comparable Market Analysis (CMA) on the house.

☞ **Money-$aving Tip #46** *Agents may admonish you that your offer is too low. Hear them out, but don't feel you have to go higher. Agents are legally bound to present all offers to the seller. They'd rather present offers that are likely to be accepted, or at least countered. But you have every right to lowball.*

Next you need to learn as much as possible about the sellers. Why are they selling? Where are they moving to? When do they have to move? Have they had other offers on the place? For how much? How long has it been on the market? You can ask the sellers questions, but you're not likely to get straight answers. If you do see the sellers during showings, you're probably better off making small talk, finding areas of common interest, and setting a positive tone for future negotiations. Take your questions to your agent. If your agent is a subagent of the seller, he or she should not divulge information that will give you a negotiating advantage. But there's no harm in asking. The more you can find out about the seller's motivations the better.

Another factor in determining your offer price is how competitive the market is. You should be paying attention to how active the market is. Are homes moving quickly at close to asking price? Or is it a buyer's market?

Finally, ask yourself how badly you want the house and if you can afford it. If some feature of the house really makes it special and unique and the market is competitive, it may be worth spending a few thousand dollars more to be sure you can get it. If you stay in the house a long time, the money won't seem like so much. Of course, this kind of attitude can be reckless. You should focus on getting a good price for the property. You always can go up if you need to. In real estate, you make your money when you buy, not when you sell.

Statistics show that buyers offer on average about 5 percent less than the asking price. But don't let statistics guide you into a knee-jerk deduction from the asking price. You have to rely on what you have learned about your market. If a property is underpriced, it is not unheard of to offer even more than the asking price. You'll still get a good deal and you're likely to beat the other buyers who are sure to pounce on it.

In addition to deciding what to offer, pick your high price, then stick to it. Don't let the negotiations get you carried away.

☞ **Money-$aving Tip #47** *First-time buyers often can be more flexible on such issues as closing dates and occupancy, especially if they have a month-to-month lease. Use your flexibility in negotiations. The seller may be concerned about how long it may take to find another home and close on it.*

Keeping the Negotiations Moving

Your offer starts the negotiations. If the offer is countered, go through the agreement carefully with the sellers or their representative. Find areas of agreement and confirm them. Where you have disagreements, look at the easiest ones first and see what you can work out. You'll be developing a spirit of optimism that agreement can be reached. Be patient throughout. Don't focus on price alone. Be creative and look for terms other than price that can serve the needs of the sellers. Keep in mind what's important to you in the agreement and what you imagine to be important to the sellers. Be willing to compromise. When you make a concession, look for one in return from the seller. If you need to counter their counter, give them something they can agree on. Shoot for win-win. Once you've come to a verbal agreement, get the signatures as soon as possible. It doesn't mean a thing without them.

How *Your Agent Can Help*

The agent can help with your purchase offer and negotiations in many ways. Exactly how depends on your agency relationship. If you have any questions on what agency is, you should review Chapter 6. The agent with whom you have been working may be representing the seller or you, or perhaps both.

An agent who is representing you will be your advocate in the negotiations. You should still be involved. Indeed, you are in charge. The agent is duty bound to do as you wish. You can confidently work with the agent as a partner in negotiating the purchase. A buyer's agent's duty is to help you get the property at the lowest possible price.

Though subagents are not your advocates in the transaction, they still can be quite helpful. The agent cannot tell you what to offer, except full asking price. However, you should be able to get general market information, including comparables, that will help you make that determination on your own.

Most agents will help you fill out the purchase agreement to the extent that the law allows. Of course the buyer's representative will give you more explicit guidance and recommend strategies. A subagent will explain the purpose of clauses and tell you the issues you should be considering. The agents will provide you with the legal description of the property. Agents also can help you draft a financing contingency because they have timely information on the mortgage market.

Commonly Asked Questions

Q. If the seller refuses my offer but doesn't counter, can I make another?

A. You can try, but it doesn't look good. The message is that the sellers don't want to deal with you. If they were still willing to talk, they could at least have made a high counteroffer. This is the risk of the lowball strategy. If you really want a home, you should make a reasonable offer. If you're looking for a steal, be prepared to lose out many times before you find one.

Q. Will I get my earnest money back if I must pull out of the deal?

A. That's a question you should ask as you're preparing the agreement. Of course, it depends on why you're pulling out. The only rationale is that a contingency wasn't fulfilled. Perhaps the inspection of the property was unsatisfactory or you could not get financing below your maximum interest rate. The wording of a contingency clause is critical for just this reason. The seller needs to feel confident that your backing out is legitimate and that you're not using the contingency as a rationalization. This is particularly true in the case of financing.

Q. What is buyer's remorse?

A. You'll know it when it hits. Buyer's remorse could strike at any time from the signing of the purchase agreement to the close or soon after. It's a tightening of the gut or sweaty palms. It's a nightmare that wakes you up. It's a scream from deep inside: "What have I done?" It all comes from second-guessing. And it's so common they even have a name for it. Handle your transaction with care and remind yourself that you have done so when buyer's remorse strikes. It shall pass.

CHAPTER 13

Preparing for a Smooth Close

Here's the day your search for a home comes to a crescendo of documents and signatures. When it's over, you've bought the house. The title has gone to you, though you may not see it happen. The big day is bound to make you jittery and excited. For first-time buyers especially, the rapid-fire "John Hancocks" and a sense of anxiety make it all a haze until the end when you ask your attorney, "You mean it's over? I own the house?!!!" You even may get the keys right then and there. You can drive to the house and jump for joy in the living room. Do cartwheels in the backyard.

What this big day is called varies across the country. We'll call it the close, or closing. You will be closing your transaction. It goes by several other names, including *escrow, settlement,* and others. The close may be handled by a title company, attorney, escrow agent, or, in some areas, real estate agents or lenders.

As much as closing practices may vary from region to region, they have some common characteristics. Someway, somehow, title will transfer from seller to buyer. This is the day of no return. You will enter the ranks of home ownership. All the financial arrangements of the transaction will be finalized and completed. There will be a full accounting of all funds, and then disbursement of the money. We've referred to closing costs throughout the book. This is the day that you'll pay them.

Any real estate agent will tell you the deal is not done until the close. To make sure that all goes over without a hitch, you need to attend to many details starting from the moment you and the sellers sign the purchase agreement. And you also should have an attorney at your side. You will be signing documents with serious obligations and legal ramifications.

Now, are you nervous enough? You may not understand everything, but you'll be surrounded by people who do. Your job is to stay on top of things and make sure the others do what they're supposed to.

Getting Organized

Appoint yourself manager of the close. Your first task is to get organized. Set up a file for the close or, better yet, a filing system. If you don't already have a daily organizer, use a small notebook to make a to-do list each day. You should start a log and record every contact regarding the home. If a notice has come in the mail or a document has gone out, jot it down. The more detailed the record the better. If you talk to someone on the phone, jot down the date, time, notes on what was discussed, and the person's name. You'll have contact with many people—your real estate agent, attorney, lender representatives, insurance agent, and more. Your log may give you important information on tracking down a problem. Moreover, it will keep you on track and help prevent problems.

The Closing Documents

There's a lot of business to take care of on closing day. What you want to avoid are surprises. The best way to ensure a smooth close is to check all items beforehand whenever possible. See that tasks have been properly completed and documents filled out accurately. Discuss the procedure with your attorney. Certain items will require a legal review. Formally request to see the closing documents 24 hours in advance of the close, which is your right.

You will get an exact tally of closing costs before the close. You will need to bring a cashier's check or money order to cover the full amount—down payment minus your earnest money deposit plus all closing costs. Don't even think about buying a house with a personal check. Bring extra cash as well, just in case of a miscalculation. You don't want to be caught empty-handed when the closing officer says, "I'm sorry, I need another five dollars to close that transaction."

Below is a partial list of the items and documents involved with the close. Some you may not see until the close. Confirm that all forms are filled out accurately before you sign.

- *Final walk-through.* Check the property a day or two before the close. All required repairs must be completed. If they haven't been or there is serious new damage, you can request compensation or that an escrow account be established to cover the cost of repairs.
- *Pest inspection or environmental inspection.* (Sometimes required by lenders.) Problems should be addressed.
- *Survey.* The lender may require a survey. Your attorney should review it for accuracy and any problems.
- *Title insurance.* Your attorney should review this.
- *Hazard or homeowners insurance.* Policy should include your name and the lender's.
- *Mortgage insurance policy.* Be sure you understand the terms and that the form is properly filled out.
- *Truth-in-Lending Disclosure Statement.* Any questions about the lender's charges or the cost of the loan should be handled well before the close.
- *Deed.* There are several types, and the purpose is to transfer the property. Your attorney should review yours.
- *Release of liens.* Any judgments or liens on the property must be paid off.
- *Mortgage or deed of trust.* This document secures the property as collateral for the loan.

- *Mortgage note or promissory note.* This is a companion document to the mortgage and describes the terms of the loan.
- *Uniform Settlement Statement.* This is a standard HUD form that itemizes all the costs of the close.

Closing Costs

You and the owners have agreed on a price for the home. You've passed the lender's scrutiny and gained approval for the loan to cover what you don't put down. You still need a bit more money. Closing costs are like a cover charge you must pay to get in the door. On average, they'll run from 3 to 6 percent of the purchase price. The costs are for fees associated with getting the loan and transferring ownership. Some are obviously important, such as title insurance or homeowners insurance; others feel more like a shakedown. It's natural to feel helpless in fighting these charges. Indeed, there often is nothing you can do about them. Still, if there are any charges you don't understand, you should ask questions. But you have to do so before the close.

In addition to your lender's charges for processing the loan, typical closing costs include title insurance, homeowners insurance, prorated real estate taxes, prorated interest, attorney's fee, survey, pest inspection, recording fee, and in some areas a property transfer tax.

The Real Estate Settlements Procedure Act (RESPA) gives some help to consumers, with its specific requirements ensuring that both buyer and seller are informed about charges. Lenders are required to provide you with the HUD booklet (or their own equivalent) "Settlement Costs and You," which is a general discussion of closing costs, including a line-by-line explanation of the Uniform Settlement Statement. You should take the time to read it over. Lenders also will provide you with a good-faith estimate of all the closing costs. They are required to do so within 72 hours of application.

You probably will be charged a prorated share of real estate taxes, adjusted in an equitable way between you and the sellers. You also will pay a prorated share of mortgage interest.

Mortgages are paid in arrears; that is, each payment covers the previous month. You will make the first month's payment at the close, and it will be adjusted to account for the time you've been in the house. This is not an extra charge, but it does mean you're putting up more cash up front. You will not need to make a payment to the lender the first month. In the second month, the regular mortgage payment cycle kicks in. If you have made arrangements for the owners to remain in the property for a time after the close, you'll be credited for the amount of the rent.

☞ **Money-$aving Tip #48** *Check all the calculations of prorated payments before the closing day.*

Title Search and Insurance

Title companies thoroughly examine the chain of title and all public records affecting the property to make sure that the seller has clear title to the property being sold. If the search shows that the title is good, the title company will issue a policy protecting the holder from any loss due to defects in the title or any liens or encumbrances on the property.

Title insurance comes in two flavors. The first is the *lender's policy,* or *mortgagee's policy.* It protects the mortgage lender only and covers losses up to the amount of the mortgage. The second, the *owner's policy,* is optional. It protects you, the new owner, in case of a flaw in the title for the house. For your own protection, you should get owner's title insurance. True, the title search for the lender's policy gives you some assurance that the title is good. If a problem is missed in the search, the lender's policy will cover the amount of the mortgage. However, you'll still be without title to your house. Owner's title insurance is well worth the cost. For a higher premium you also can purchase extended coverage, which gives you added protection for such problems as a bad survey or unrecorded liens.

Who pays for title insurance varies in different parts of the country. It's logical that the seller would be responsible for confirming that the title is clear. Sometimes sellers pay for the title search and buyers pay for the insurance portion. The costs may be negotiated, or the custom may be that the buyer pays all. You should ask your real estate agent about local practices.

The escrow company or lender may have a favorite company because of a negotiated agreement or outright kickbacks. Any explicit relationship should be disclosed to you as required by RESPA. Fees average about ½ percent of the cost of the house. Ask the lender about using another title company if charges are extraordinarily high. If you're buying the policy, you have the right to shop around. In some parts of the country, real estate agents or lawyers serve as agents for title companies.

Homeowners Insurance

Your lender will require that you purchase hazard insurance, which protects the property from physical damage due to such hazards as fires and storms. You should purchase a broader homeowners insurance policy, which includes coverage for theft and liability. Homeowners policies vary as to what is covered.

You should buy a policy that provides for replacement value. Your insurer will then pay for the cost of offering the same functional utility but using current materials and techniques. You should be clear on how a policy defines replacement value and what would happen in the worst-case scenario. For instance, rebuilding an older home after a fire will require that you build according to the current code, which may include features that were not in place in the existing structure. What will your policy cover?

The typical homeowners policy does not include flood or earthquake coverage. If your home is located within a flood plain, your lender will require that you purchase flood insurance. The premium is a set amount nationwide.

☞ **Money-$aving Tip #49** *Look into homeowners policies from your auto insurer. Some companies offer discounts for owners of multiple policies.*

Shop around your area for insurance companies. Ask around to see who other people are using and how much premiums run. You can check the financial stability of insurance companies through A.M. Best or Standard & Poors. Personal finance magazines like *Money* or *Kiplingers* regularly run features on buying insurance. Be sure to bring your proof of insurance to the close.

After the Close

You should do a bit of follow-up after the close. Most important is to make sure that the deed is properly recorded. Be sure to follow any other instructions regarding the transfer of documents. When all is done, make photocopies of all the documents. File copies in a safe place.

If you're planning projects for the new home like painting, new carpeting, or small remodeling jobs, it's best to take care of them in advance, before you have to move your belongings around. You should at least allow a full day for cleaning before you move in. It's also good to give the house a security check, making sure window and door locks are sufficient and in good working order.

Next you have to get ready for the move. You probably know a bit about all the details involved with moving. If you're on the ball, you've already made some of the arrangements. You'll need to contact the phone and utility companies and fill out a change of address card at the post office. Is the new home cable ready? You'll want your MTV. Now's also a good time to have a garage sale. Small comfort after spending your life's savings, sure, but you can make a few bucks and clear out your junk. Don't move any more than you have to.

If you're like most people, you dread moving. Look at it this way. Moving will never get any closer to fun than when you go to your first home.

How *Your Agent Can Help*

This chapter hasn't gone into great detail about what happens at the close because so much of it depends on local laws and customs. Once the sellers have officially accepted your offer you should request a short meeting with the real estate agent.

Ask the agent to describe how the close will be conducted and what fees you'll have to pay. The weeks leading up to the closing date will be a busy time. Your contract will specify when that closing date is and also will define time frames in which you must meet contingency conditions. These deadlines are critical. Not meeting them can cause the deal to fall apart. The two tasks you'll want to get right on are arranging for a home inspection and applying for the loan. Real estate agents often will offer substantial assistance with your loan application. The loan application process itself is full of details you should follow up on. For example, you must be sure your employer cooperates in getting documentation to the lender in a timely manner. The agent should make sure the lender's appraiser gets into the house okay.

Your goal in the meeting is a full understanding of what happens at the close. Agents will often do a lot of follow-up in preparation for the close. They may not have as much riding on that big day as you, but it is nonetheless very much in their interest to prepare. You should know what you need to do, what the agent will do, and what your attorney needs to do. You should ask what steps the agent will follow up on and what you should be following up on. Then you should follow up on your agent's following up. Remember all the while that you are dealing with busy people and yours is not the only transaction. That's why confirmation is so important. Of course, take notes during your meeting.

Commonly Asked Questions

Q. My lease is coming up before the close. Can I move into the house early and pay rent to the sellers?

A. The deal is not done until the close. Therefore, the seller is unlikely to let you into the property early. Certainly, the seller's attorney would advise against it. The same goes for starting any improvement or rehab projects. Bad enough if the deal falls through without having to remove would-be buyers from the house or deal with half-completed projects. You're better off trying to work something out with your landlord.

Avoid this situation by giving yourself a cushion when you make arrangements with the landlord for leaving. A variety of factors can lead to a change in the close date. If you get stuck and the sellers actually do let you move in early, make sure your insurance is in force. Liability starts the day you move in.

Q. What is an abstract of title?

A. Abstract of title is a condensed history of all records affecting the property going back to the origination of the title. The abstract does not insure the validity of the title. It merely states the public record. It is essentially a type of title search used in some states. In no way is an abstract replacement for title insurance. Your lawyer should review the abstract for accuracy.

Q. Can I see a copy of the house's appraisal?

A. You paid for the appraisal, and you're entitled to a copy. Your lender may or may not volunteer one. If there is any resistance to getting a copy, put your request in writing. Lenders are required to give it to you.

CHAPTER 14

Managing and Maintaining Your Investment

Congratulations! You're in. You've bought your first home. Odds are you've made the biggest investment of your life also. You don't need to be told how important it is to protect your investment. In fact, you may be anxious about it. Perhaps you still haven't shaken off the last shreds of buyer's remorse.

Throughout this book we've emphasized that amidst all the dollars and cents of homebuying are lifestyle concerns. Moving into a new home is exciting because it's the beginning of a new life, especially so when you become a homeowner for the first time. However, just as you bought the home in a business-like manner, you need to take care of business during your period of ownership.

In Chapter 1, we mentioned tax deductions among the benefits of owning a home. You may be accustomed to using the IRS's 1040EZ form. You now will be itemizing your deductions. You should start getting ready right after the close, filing away your settlement statement for future reference. Good tax planning always begins on January 1 of the tax year. Not in December, when you're scrambling in search of deductions, and certainly not in the next year, when all you can do is gather up your documentation. In this chapter, we'll tip you off to the common deductions for homeowners. However, don't use this advice as a substitute for the professional guidance of an accountant or professional tax preparer. The detailed guidelines of the IRS are full of exceptions and qualifications. As of this writing, our tax code could not be called simple. It is impossible to cover tax issues in the necessary detail here. Even if it were possible, tax laws are always subject to change.

To protect your investment, you also should practice preventive maintenance on your home. Look around the house every spring and make a list of summer projects. Check again in the fall to be sure you're ready for winter's elements.

On average, first-time homeowners sell in about seven years. That might seem like a long way off, but it's not. As you begin to make modifications and improvements to the home, keep the perspective of a would-be buyer. You'll be selling before you know it.

Annual Income Tax Deductions

Mortgage interest and real estate taxes are deductible. Your lender will send you a statement early in the year stating the amount of interest you paid for the year. Note that only the interest is deductible, not your entire mortgage payment. If you decide to get a home equity loan later on, that interest also will be tax deductible. Interest on any loan for which your house is the security is tax deductible. You can deduct interest on mortgage loans totaling $1 million. Again, IRS guidelines are subject to change and mortgage interest is one deduction Congress has been looking at recently. We can hope that any changes won't affect deductions for the size of mortgages typically taken by first-time buyers.

You should keep your mortgage statements for the year on file. If the interest amount the lender supplies doesn't look right, go back and total up the amounts on your statements.

You'll get an extra deduction in the year you buy the house. You can deduct the points the lender charges for the loan. Points also may be deductible on home equity loans or refinancing, amortized over the life of the loan. Consult an accountant or the IRS to get the guidelines. Origination fees and other charges of the bank are not deductible. However, you can add them along with all other nondeductible closing expenses to your basis, which will be discussed below.

Your lender probably pays your real estate tax bill for you, collecting money each month through your escrow payment. Be sure that you deduct only the real estate taxes, not the full amount of your escrow payment. Taxing bodies often send real estate tax bills to the homeowners and the lenders. If you're not getting a copy of your real estate tax bill, ask your lender to send a copy. Use the tax bill as your record in making the deduction. Note that special assessments for an improvement to the neighborhood or block are not tax deductible. But as with home improvements, you can add them to your basis.

Unless you're self-employed, a certain amount of money is withheld from each paycheck to go toward your end-of-the-year tax obligation. How much depends on the withholding allowances you claim on your W-4 form. With your new deductions, your tax bill will be less. At the end of the year you should get a bigger refund. Instead of "lending" your money out to the IRS, you can change your withholding allowances and take home a slightly bigger paycheck. Contact your human resources department to get the form. On the back is a worksheet to help you calculate your withholding allowance. If you take too much, you may run into penalties or interest charges at the end of the year. The safest way is to consult an accountant. Or just be conservative and increase it later if you still are getting a substantial refund at the end of the year.

The quarterly payments required by the self-employed serve the same purpose as withholding tax, assuring that you pay taxes gradually over the course of the year. Consult an accountant as to whether or not you should reduce payments. The answer depends not only on the mortgage interest deduction, but also many other factors in the business.

☞ **Money-$aving Tip #50** *Generally, moving expenses are not deductible. However, if your move is job related and you've moved 50 miles or more, they might be. See IRS Publication 521 for more information on requirements, eligible deductions, and dollar limits. Or consult an accountant.*

Property Assessments

Even though real estate taxes are deductible, they're a big bite from your wallet and you don't want to be paying more than your fair share. We have advised you to look at typical real estate tax rates in a neighborhood as part of the home-hunting process. You can't assume, however, that your rate will be the same as your seller's. How much you pay in real estate tax depends on the *assessment* of your home. The assessment is a valuation on your home as determined by local government officials. In some localities homes are reassessed after a sale, which means you'll probably get a tax hike unless you paid less than your seller did. Local officials generally assess properties in the neighborhood on a regular cycle. You can ask your neighbors when the next assessment will be.

Fortunately, you do have some recourse if you feel you are paying too much. First you should look at the tax bill and make sure that the information is correct. If your property is listed as having four bedrooms when you actually have three, you're in luck. Even if the information is correct, you can argue that they valued the property too much by comparing it to the valuation of other houses on your block. The valuations are often published in local papers. The exact process for appealing your property taxes varies from state to state. You should contact the local assessor's office. Believe it or not, assessors can be quite helpful. Some even publish information on how to challenge your assessment.

Your Basis

When you sell the home, the profit you make is subject to taxation as a capital gain. Profit is determined by subtracting your *basis* from your selling price minus selling costs, such as the commission you pay the agent. Basis is the dollar amount the IRS attributes to an asset when it calculates the capital gain or loss.

Basis = Price + Purchase expenses + Improvements

The basis of your house is the price you paid plus any non-deductible expenses associated with the purchase; that is, probably all your closing costs with the exception of points. This is just one reason to keep your closing documents safely on file. You also can increase your basis by adding in money you've spent on improvements to the property. What constitutes an improvement will be discussed later. You want your basis to be as high as possible, so as to avoid the capital gains tax. The total after additions and subtractions is called the *adjusted basis.*

Calculating basis can get complicated. For instance, you must subtract any depreciation you've taken on the property. Personal residences normally are not depreciable. However, you might be depreciating business use of the home, such as for a home office.

Rollover

You probably already have pen in hand to write your congressman a letter on how you now support a reduction in the capital gains tax rate. While that may be a good idea, there's no rush. The good news is that Uncle Sam has another break for homeowners–the rollover provision. With the rollover you still keep the tax liability, but you delay paying it if you buy another home of equal or greater value within two years of your sale; that is, if you "roll over" your profit into another investment.

As a simplified example, ignoring home improvements, closing costs, and selling expenses, suppose you buy a home for $100,000 and later sell it for $120,000. Your profit is $20,000. You roll over that profit to buy your next home, which costs $150,000. The rollover reduces your basis each time you sell. Because you rolled over your profit of $20,000, your adjusted basis for the second home is $130,000 (the cost of the home minus your $20,000 profit). Five years down the line you sell that home for $175,000, your profit is now $45,000. You can roll that into the next home as long as you spend $175,000 or more.

☞ **Money-$aving Tip #51** *In order to qualify for the rollover, both the original house and the replacement house must be your principal residence. Vacation homes don't apply. You will surrender this benefit if you convert your home to a rental property, though renting for a short time prior to the sale is allowed. Consult an accountant for the details.*

The process can be repeated indefinitely, as often as once every two years. You can see what a large amount of untaxed profits you are accumulating. Eventually, though, the bill comes due. But Uncle Sam delivers yet another goodie—the one-time exclusion. There are limitations. You must be 55 or older and have owned and occupied the home as a principal residence for three of the five years before the sale. You may use the exclusion only once, and you are not eligible if you're married to someone who has already used it (known by accountants as a tainted spouse). If you satisfy the requirements, the first $125,000 of capital gain is tax free. How about that! If you're in a 28 percent bracket, you'll save $35,000 in taxes.

Home Repairs and Home Improvements

What's the difference between repairs and improvements? Repairs are not tax deductible. Neither are improvements; however, you can add them to your basis. From a tax standpoint improvements are definitely better. So what qualifies?

Improvements should add to the property's value. An improvement is something new and different you've done to the property. Repairs, redecorating, and maintenance are not improvements and therefore not tax deductible. Suppose you have a leak in your roof. You call in a roofer, who patches it up. That's a repair. Suppose the roofer warns you that the roof is old, and you can expect more leaking. You decide to reshingle the whole roof. That's an improvement. Painting the walls is not an improvement. But if you're painting the walls of an addition you've put on, that is an improvement. Replacing a broken window pane is not an improvement. Replacing the whole window unit is. You get the idea. Surely there are gray areas so consult an accountant. In certain instances, it may come down to how aggressive you want to be on your taxes.

☞ **Money-Saving Tip #52** *Rehabbing a property is considered entirely an improvement. All related expenses may be added to your basis.*

Recordkeeping

By now you've realized that it's important to keep good records. As mentioned, you should have a set of your closing documents in a safe place. Important documents like the title and mortgage are best put in a safe deposit box. You'll need copies of all closing documents in your house for easy reference. The settlement statement is a handy resource because it summarizes all the costs of the close. You can use it to see the total you paid in points. You'll also want to keep it around for when it's time to calculate your basis.

You should keep your tax returns also. The IRS sets a period of three years for its routine audits. However, if it suspects that you have underreported income by 25 percent or more, it can request records from as long as six years back. If fraud is suspected, there is no limit. It's therefore a good idea to keep your income tax report and all supporting documentation for a period of at least seven years. It's safest, of course, to keep everything. You should at the very least keep the tax filing from the year you bought the house. It will be backup of sorts for the closing documents and also will be a useful reference as to how you reported costs of acquisition.

You'll want to maintain a file of all home improvement receipts. Any documents of transactions that will affect your basis should be kept on file.

Though there are no tax ramifications, you should keep receipts of major ticket items for insurance purposes. The best way to make sure you're fully compensated in case of theft or fire is to take an inventory of all items of significant value in your house. Photographs or videotape is another way to document your property. Of course, it's best to keep these records outside of the house.

Preventive Maintenance

You'll have plenty of home projects when you buy a house. Some people love to tinker around the house fiddling with this and that. Others dread handy work. Whichever camp you fall in, don't ignore preventive maintenance. It's easier to call a plumber to make a repair before tragedy strikes. Call the plumber on a weekend as water drips down your wall and you're looking at double time. You don't want to pay those rates.

The biggest threat to the structural integrity of your house is water. Excessive moisture can cause peeling paint, discoloration, wood rot, masonry cracking, and many other problems. Check the grading around your house every year to make sure that water is steered away. Make sure your gutters are clean and water is directed away from the house. Water also can be a problem from the inside. Your bathroom has a ventilation fan, and you should use it. Ideally, the kitchen should be ventilated also. If there's no fan, open a window when you're cooking. Monitor your humidity levels and use humidifiers or dehumidifiers, if necessary. Be on the lookout for signs of moisture in crawl spaces, basements, and attics.

You should inspect the roof at least once a year. Patch cracks with plastic cement. Trim tree branches so they don't brush against the roof, even if loaded down with snow or ice. Make sure the chimney flu and vents are clear of obstructions.

Check that doors and windows are properly weather stripped. In northern climates you should have storm windows or double-pane windows. Otherwise condensation will build up on the inside window, possibly leading to moisture problems.

Be sure you understand the basics of the electrical and plumbing systems. You should take notes during the inspection. Sometimes owners can be quite helpful showing you around during the final walk-through. Know where the electric panel is as well as the shut-off valves for the main water supply line, plumbing fixtures, and gas lines. If you need to shut down any of the mechanical systems in an emergency, you will be in a hurry. You don't want to lose a minute looking around. If there is a septic tank, you should know where it is and what the maintenance procedures are.

Visit your bookstore and see what they have for home maintenance. A homeowner's bookshelf should have a few good reference books to get you started on repairs. An inexpensive resource for homeowners is the University of Illinois School of Architecture-Building Research Council. It publishes a dozen or so pamphlets, for just a few dollars each, on such subjects as energy, moisture, windows, roofing, and planning. Call 800-366-0616 for a listing of its publications.

Remodeling

Now that you own your home, you can do whatever you like. You are king of your castle.

But wait. Unless you plan to hold on to the castle the rest of your life, you should consider the next people and the fact that you will one day be selling. Certainly you should give expression to your tastes when decorating a house. However, don't indulge any particularly esoteric passions that will be difficult for the next people to undo. For instance, a Japanese rock garden might be your idea of a great backyard. And maybe you'll be able to pull it off. But you might be better off trying to find one around town that's open to the public and keeping your lawn intact. Think twice about any improvement. Even some that seem undeniable in your mind may cause future buyers to pause. Your vision of a built-in swimming pool might be a liability risk and maintenance money-hole in a future buyer's eyes.

If you have the time and inclination, you should try to do whatever you can yourself around the house. You can save a lot of money. But you should learn to distinguish a good do-it-yourselfer from projects that are better off handed to a pro. A perfect example is repairing seriously damaged walls. In one sense, it's fairly easy. Just mix up your "mud" and glob it on. However, to do it well requires experience. It's worth it to pay somebody else and be able to appreciate a higher level of craftsmanship. On the other hand, if you're patching a hole under the sink, go for it!

FIGURE 14.1 Remodeling projects and how they pay off

Type of Improvement	Recovery Cost (Percentage)
1. Room addition	70–90
2. Major kitchen remodeling	45–70
3. Minor kitchen remodeling	60–80
4. New bath	75–100
5. Bathroom remodeling	60–80
6. Master suite	60–80
7. Reroofing	10–30
8. Finished basement	30–45
9. Garage	30–50
10. Windows and doors	25–45
11. Insulation	0–25
12. New heating system	30–45
13. Deck	65–75
14. Sunspace	5–20
15. Swimming pool	0–65+ (varies by market)
16. Skylight	0–30
17. Exterior painting	40–50
18. Siding	15–35
19. Landscaping	45–65
20. Energy-efficient fireplace	45–100

Remodeling is a good way to improve your living comfort and add value to your house at the same time. The chart in Figure 14.1 shows the typical return on the dollar for home improvements. However, watch out for overimprovement. After a certain point you will cease to get a good return on your dollar. You don't want to have the most expensive house on the block. Remember that buyers will be looking at comparables, your neighbors' sales prices, before they decide what to bid on your house. You also should be sure to get any required building permits. Many property disclosure laws require sellers to indicate if building permits were obtained for improvements.

How *Your Agent Can Help*

Once you own a home, you'll find that you become very interested in the selling price of homes around the neighborhood. To some, it becomes an obsession. So it was in the heady days of the 1980s when homes appreciated in double digits every year in many parts of the country. Cocktail party chatter was all about who got what for their house and how quickly it sold. People are a little quieter when times aren't so good. You don't want to get carried away in all this, counting your profit on a regular basis. The profit is only on paper until you make a sale, and market value on your home isn't truly determined until you and a new buyer shake hands.

Nonetheless, it's sensible to periodically track your investment. If you are curious about what other homes in your area are selling for, feel free to call your agent. Most agents will be pleased to answer your questions. They want happy customers to stay that way. It could mean referrals or another call for comparables when you're ready to sell.

Commonly Asked Questions

Q. *Is it a good idea to prepay your mortgage to save money on interest?*

A. The answer depends mostly on your philosophy of debt. Mortgage interest rates are generally low when compared to any other source of credit. And they are fully tax deductible. As debt goes, mortgage loans aren't so bad. You could take pencil to paper and make a comparison to what you're saving by accelerating payment and what you might make on other investment opportunities. It's probably safe to say that prepaying your mortgage should not take precedence over other savings plans. You'll want to build a nest egg for other purchases. You won't get the money back in home equity until you sell.

On the other hand, for people with spending problems, pre-paying the mortgage may be the best way to save money.

Q. How can I keep my energy bills down?

A. One place to start is using energy wisely. Turn down your heat at night before you go to bed. If your house is empty during the day, keep it low then. Similarly, don't bother with the air conditioner if nobody is home. Consider the effect of the sun. Keeping a shade closed when the summer sun is shining through will help keep your house cool. A well-placed tree will add energy efficiency to the home as well as beautify the property. In the summer months, it will shade the house. In the winter, the leaves will be gone and the warming sun will shine through.

A major source of heat loss is windows. Inspect them carefully for drafts during the winter and weatherstrip appropriately. The next best place to look for savings is the roof and attic. Make sure your insulation is adequate.

Q. Can I add the value of my personal time in a home improvement project to my cost basis?

A. No. If you do the work on an improvement you can add the cost of materials and equipment to the basis, but you cannot charge for your own time.

CHAPTER 15

From the Desktop to Cyberspace

Technology may have been slow in hitting real estate, but now the industry is making up for lost time. A driving force has been a court decision declaring that the Multiple Listing Service (MLS) system of real estate boards would no longer have the exclusive right to listing information on homes. The decision brought new competition, opening the door for database and communications companies or anybody else to deliver the listing information.

The change also has inspired a higher level of customer service in the real estate industry. No longer can brokers count on buyers and sellers going to them as the only source of property information. Now, real estate agents have to focus on their added value to the transaction, their expertise and advice. They also have upped the ante on the technology. The National Association of REALTORS® and other major vendors have formed an industry consortium to develop a nationwide network of listing information.

Real estate agents use technology heavily in their marketing and to communicate with customers. Many real estate professionals and assorted marketing companies are creating sites on the World Wide Web. For the homebuyer with a computer and modem, it means loads of free information. In this chapter, we'll list some of the best sources. We've done our best to select those that are likely to be around for a while. But if you've done any Web surfing at all, you know how quickly the Internet is changing. Some of the sites in this chapter already may have changed URLs (universal resource locators—the "addresses" of the Internet). It's a good bet that the information on the site will change. The capability for easily updating and enhancing information is a key advantage of the Web. With a working knowledge of Web indexes and search engines, you'll be able to find cool and useful sites on your own. The Web is a perfect companion to the information in this book. Not only can you get updates on the topics covered here, you also can search out local information.

If you don't have Web access at home or at work, you may still be able to access this information. More and more public libraries are now offering Internet access. Ask at your local library if it offers this service. If so, you'll probably have to reserve a time. Come organized with your list of URLs, because time will fly when you're on the Web. Go to the library early and tell the reference librarian what you're looking for. The library may have print resources that will fit the bill, and you can devote your Web activity to the most time-sensitive information.

What the Pros Are Doing

Real estate has always been a people business. Today's technology makes it easier for real estate agents to communicate with customers, sellers, loan companies, and other professionals. Agents are almost always out of the office, but with voice mail, pagers, and cellular phones, they're never far from their customers. From their desktop computers, agents can deliver all kinds of information to you. Not so many years ago, MLS information on homes for sale was kept in big fat books. Today, though, most offices have had computer access to the MLS database for some time. Now there's all kinds of other information available online. Your agent may be able to get you property tax data or school information through online databases.

Loan companies are using technology in the application process. The loan approval process has been streamlined so that approvals that once took a few weeks can now be accomplished in a few days. It's still not common to actually apply for a loan via computer, but the day when that is standard procedure is probably not far off.

Real estate brokers, lenders, and allied professionals of all sorts are posting their information on the Web. Look for the URLs in the ads of your local real estate brokerages. Jot them down and do a bit of shopping via computers. You'll find listings in your area and other useful information. Some sites are purely promotional, others offer rich information along with the sales pitch. Many banks, for instance, are putting personal finance information on their Web sites.

Software

The best software investment you can make for your home hunt is in the personal finance software. The most popular program is Quicken, which is available for both Macintosh and Windows-based computers. Quicken allows you to track and categorize all your transactions. It is easy to run reports that will give you a clear idea of where all your money is going. First-time buyers looking for ways to tighten their belts and save up for a down payment will find this to be a big help. You can start saving by setting up a budget. The program is a real time-saver in organizing your records for tax time. It is compatible with the tax preparation programs TurboTax for Windows and Macintax for the Macintosh, so transferring your records is seamless. Quicken also will amortize and track loans. You can use this function to compare what your payments would be for various loan packages. Another popular program is Microsoft Money for Windows.

If you have a CD-ROM drive, you can buy programs that will help in home repair and improvement projects. Books That Work produces quality CD-ROMS on home design, home improvement, and landscaping. Its Web site is described below. Check computer magazines, mail order catalogs, and software shelves for similar products. The multimedia capabilities of CD-ROM are ideal for home repair and rehab work.

Dearborn Financial Publishing has come out with a line of book/CD-ROM packages based on its popular business books. Real estate topics include *The Home Inspection Kit, The Mortgage Kit, The Homebuyer's Kit,* and *The Homeowner's Kit.*

Online Services

If you're serious about the Internet and Web surfing, the best way to do it is with a direct Internet connection through a local Internet access provider. However, such access may be hard to find in rural areas and small towns. Also, these services demand a bit more technical expertise on your part. You'll have to install software or the connection and perhaps do some configuring.

The next best way to access the Internet is through one of the major consumer online companies, such as America Online (AOL), CompuServe, Microsoft Network, or Prodigy. In addition to Internet access, these services offer proprietary information not available over the Internet, including message boards and chat areas for communicating with other members. If you already subscribe to one of these services, you should dig around for real estate information. Prodigy features in-depth information in corporate-sponsored, premium areas. Compuserve is especially known for its databases of newspapers and magazines.

If you don't have an online account but are looking to get started soon, you won't do better than AOL for real estate information. (If you're already a member, use the keyword "Real Estate.") Peter Miller, author of *The Mortgage Hunter* and several other real estate books, runs the Real Estate Center. You'll find message boards on dozens of subjects of interest to professionals and consumers alike. You also can access listings and mortgage rates from around the country. You can download freeware and shareware software programs, such as amortization software and loan qualifier programs. Shareware is software from small developers who ask for only a nominal fee. Freeware is free. You can download the programs for free, so payment depends on the honor system. You should pay for any shareware you use to encourage this kind of development. Also available are text files that are excerpts from books, articles from experts, or press releases from major real estate companies or government agencies. The Real Estate Center also features direct links to some of the best real estate information on the Web. You'll find other real estate information on AOL. HouseNet provides information on home improvement projects. The Magazine Newsstand features areas for such magazines as *Home, Worth, Consumer Reports,* and *Business Week.* The Homeowners Forum offers a chance to discuss issues from financing to repair. And you can visit CENTURY 21® Communities, which will be described later.

World Wide Web Resources

The World Wide Web offers an astounding array of resources that is growing all the time. Unlike other media, much of the information comes to you unfiltered. Sites may go up without the benefit of editors, publishers, or proofreaders. You need to be a little more critical of these resources. Remember the old adage, "Don't believe everything you read in the paper" ? Well, it certainly holds true on the Web. Take note of what the source of the information is. Is it authoritative? Is it objective? Government information sources are very reliable, and often data-rich. Commercial sites come with a point of view. Not that the information is unreliable, but it might be slanted. Well-designed pages are dated, so you can note how current the information is. You also can make some judgments about the quality of the content by its presentation. Is the site well-organized and easy to navigate? Are the pages well written and free of typos? Are graphics useful, and not so big that moving around the site takes forever?

We've included search engine and index sites in this list. The Web is always changing, and search engines allow you to stay on top of it. You can use them to find sites local to your area. Real estate brokerages and agents, lenders, and local newspaper sites are all potential sources of information.

Homebuying Resources

Here are some favorites for all-around homebuying advice. We strongly recommend visiting the International Real Estate Directory as a starting point.

Homebuyer's Fair
http://www.homefair.com
Homebuyer's Fair is a private company founded by Arnold Kling, a former economist with the Federal Reserve Board and Freddie Mac. This is a diverse information source and includes a first-time buyer section. The site features interactive software to help with mortgages and relocation.

House Hunter's Helper
http://www.hypervigilance.com
This whimsical site offers a range of information and includes a handy form for taking notes while you view homes. Just print it out, make photocopies, and take it along. Note that this is not a real estate site. If the form is removed, you'll be out of luck. On the other hand, there are enough creative ideas here to make the trip still worthwhile.

Inman News Features
http://www.inman.com
Prepared by a team of real estate journalists led by Oakland-based real estate journalist Brad Inman, this site is a great source for the latest news on homebuying issues.

International Real Estate Directory
http://www.ired.com
The tagline for this site, "For everything that's real estate," lives up to its promise. The first stop for home hunters in cyberspace, this site features a searchable directory with ratings of thousands of real estate-related sites. You can look up financing and home information specific to your region or search for brokers or agents in your area. The site was created by Becky Swann, a Texas real estate agent active in buyer's agent organizations. The site includes news items from Oakland-based real estate journalist Brad Inman's *Inman News Service,* which has its own Web site.

RTK NET
http://www.rtk.net
The Right-to-Know Network provides free online access to over 100 gigabytes of quantitative databases and numerous text files and conferences on the environment, housing, and sustainable development. You will need to use Telnet to reach some of the databases. Two nonprofit organizations, OMB Watch and The Unison Institute, have assembled this incredible resource, much of it derived from government information.

SchoolMatch
http://ppshost.schoolmatch.com/oldindx.htm
SchoolMatch is a school research, consulting, and database firm. The site includes a glossary of terms, national statistics on K-12 schools, and news about education. You can order reports for specific schools for a fee. Sample reports are provided. The site serves parents, REALTORS®, homebuyers, and lawyers. The file name for the home page, "oldindx.htm," leads one to think the URL might change (olderindx.htm?). Public Priority Systems, which provides the content for the site, offers several related services, including litigation and child custody consulting. Even if the SchoolMatch home page moves, you'll find it by digging around the site.

Texas A&M Real Estate Center
http://recenter.tamu.edu
This is an especially good resource for Texans. The audience includes professionals and the site covers commercial real estate interests as well. However, the information is authoritative and worth the visit even for homebuyers who don't live in Texas.

Credit and Loans

Financing is one area that's perfectly suited to Web publishing, and these sites are timely and interactive. Many sites include forms or interactive calculators to help you crunch numbers. Rates change daily, so the Web is a good way to check the latest trends. We're listing quite a few here and still weeded some good ones out.

HSH Associates
http://www.hsh.com
This site provides a major source of home mortgage information. The Author's Corner section includes articles and government publications on a range of homebuying issues.

Countrywide Home Loans, Inc.
http://www.countrywide.com
Countrywide is the largest lender of mortgages. In a business that's very local, it's one of the few companies that's active nationwide. The site contains a glossary and interactive prequalification forms.

Fannie Mae
http://www.fanniemae.com
The Federal National Mortgage Association, or Fannie Mae, is a huge and influential player in the mortgage industry. Here you can learn about new mortgage information from the people who set the rules. This site also includes related reports and consumer information.

Finance Center
http://www.financenter.com
Finance Center is a Web-based business that helps consumers with personal finance information on homes, boats, autos, and credit cards. Interactive calculators are excellent. It will hook you up with "best rate" service providers, but only uses a couple of sources for homes.

HomeOwner's Finance Center
http://www.internet-is.com
To access this excellent mortgage site, go first to the Internet Information Systems home page and look for the link to HomeOwner's Finance Center, which is a California mortgage brokerage firm that can make loans in 43 states. The site includes background information and tracking records of various indices used in adjustable-rate mortgages. You can apply for a loan online with a rapid approval turnaround of one business day.

InfoBank—Bank Rate Monitor
http://www.bank-rate.com

Bank Rate Monitor is a publisher that tracks bank rates. InfoBank serves consumers with general personal finance information and is an excellent source for late-breaking mortgage news. This site has a speedy loan calculator for evaluating various payment and loan options.

Mortgage Market Information Services
http://www.interest.com

Mortgage Market Information Services tracks mortgage rates nationwide for newspapers. Get it here before you read it in the Sunday paper. You can get rates by state and the latest mortgage news from a variety of sources. The site includes a glossary of mortgage terms for consumers.

Mortgage Mart
http://www.mortgagemart.com

This site contains great step-by-step information on the mortgage application process. The mortgage library features articles from industry experts, federal regulations, a lending overview, definitions, amortization schedules, and a rent-buy calculator. The Locator will help you search for mortgage professionals by city and state. Inland Mortgage is the major sponsor of the site, though it is unclear who is responsible for the site's content.

Government Information

You can count on authoritative information from these sites. You can go right to the source for information on FHA and VA loans. If you'd like to invest in foreclosures, you'll find listings on the HUD site.

Government Services Administration (GSA)—Consumer Information
http://www.pueblo.gsa.gov

The GSA offers consumers inexpensive information pamphlets on housing issues. This site includes a catalog of offerings, all for less than a dollar, some free. (The GSA home page is http://www.gsa.gov.)

Housing and Urban Development (HUD)
http://www.hud.gov
A superb consumer site from the government agency, this site is packed with useful information and links, especially to other government information resources, though a bit slow because of the graphics. (You can counteract this by setting your browser to not load graphics.) It includes listings for homes from HUD and the VA, and lots of information on options for low-income buyers, FHA loan programs, environmental issues, statistics, and demographics.

VA Loans
http://www.va.gov/vas/loan/index.htm
The Department of Veterans Affairs provides information on VA loan programs at the above site. If you don't find what you need at the above URL, go to the VA home page: http://www.va.gov.

Associations

Below is the site from the National Association of REALTORS® and two state sites that are especially good. The REALTOR® Information Network links to state and regional boards around the country. Check for home pages in your area.

California Association of REALTORS®
http://www.car.org
This is the largest of the state associations. This site serves its REALTOR® members, but has useful information for California homebuyers. You can get state demographic and economic information or link to member's Web pages.

Northern Virginia Association of REALTORS®
http://www.nvar.com
This is designed for REALTOR® members. Check the news section for timely information. You also can reference members' Web pages.

REALTOR® *Information Network*
http://www.realtor.com
The official site of the National Association of REALTORS®, the REALTOR® Information Network offers listing information as well as consumer advice. The site also includes links to state and regional boards.

Listings

We've focused on big sites with nationwide coverage or plans for it. Don't overlook the Web sites of your local real estate brokerages. The choices for listing information on the Web is sure to grow quickly. You also can look for listings in the online pages of your local newspaper.

Home and Land Magazine
http://www.homes.com
Home and Land publishes 250 print magazines in 43 states and features a large inventory of listings. It also has software and other features.

HomeScout
http://homescout.com
Homescout is from an Internet marketing firm called The Cobalt Group. It works like an Internet search engine for real estate listings, storing summaries of a few hundred thousand listings in its database. Search for homes via city, state, proximity, price, and number of bedrooms and baths. Links are provided to community home pages, attorneys, insurance companies, and other home-related information.

The Living Network
http://usa.living.net
This listing site is owned and operated by the Florida Association of REALTORS®. Its aim is to be nationwide.

Condominiums and Cooperatives

We found a useful site specializing in cooperative and condominium association issues. Look around for information in general real estate sites as well.

New York Cooperator
http://www.cooperator.com
The New York Cooperator is a magazine serving the co-op and condo community. The site provides information on co-op and condo ownership issues and has a few articles from the current issue. You may not find exactly what you're looking for in one visit, but you can get a feel for the issues of condominium or cooperative living.

House Design and Improvements

If you're looking for a fixer-upper, check out these sites. You'll want to revisit them after you buy your home.

Books That Work
http://www.btw.com
Books That Work is a publisher of CD-ROMs on homes and gardens. The site features samplers from their CD-ROMs and more for homeowners and homebuyers. However, if you're building or buying a new home, you should check it out. The site has good tips for all buyers on what to look for in design.

HouseNet
http://www.housenet.com
From syndicated remodeling columnists Gene and Katie Hamilton, this site features articles on home improvement projects, tips, and a message board for exchanging ideas with other do-it-yourselfers. One section includes an extensive lists of articles to help you decide whether to do a project yourself or to hire out. The site started in 1991 as a bulletin board system (BBS) and boasts a long list of press clippings singing its praises.

Home and Design Online
http://www.hdo.com/chat3.html
On this site you can "chat" with other participants on a range of home-related issues. Some discussions are sponsored by vendors selling products or services. It also links to other real estate chat sites.

Personal Finance

Sound Money
http://money.mpr.org
"Sound Money" is a weekly radio show produced by Minnesota Public Radio. The site features information and resources on a range of personal finance issues. The Found Money section is a message board of ideas for frugal living. Check it out while you're scraping up your down payment, then go back and tell 'em how you did it.

Search Engines

The following are the major search services at the time of this writing. They use different technologies and interfaces. Each has strengths and weaknesses. Some also index the Web. These sites will evolve and new ones will come along. Try them out and see what you like.

Excite
http://www.excite.com

Magellan
http://www.mckinley.com

Point
http://www.point.com

Inktomi
http://inktomi.berkeley.edu

Alta Vista
http://altavista.digital.com

InfoSeek
http://guide.infoseek.com

Lycos
http://www.lycos.com

Open Text Index
http://www.opentext.com/omw/f-omw.html

WebCrawler
http://Webcrawler.com

Yahoo
http://www.yahoo.com

Usenet News Groups

Usenet newsgroups are ongoing online discussions on a common topic of interest. Some 20,000 or more newsgroups are on the Internet on just about every subject imaginable, including real estate. Newsgroups differ from e-mail discussion lists, also known as listservs. To read e-mail discussion lists, you subscribe and postings automatically go to your e-mail address. Newsgroups work more like a bulletin board. To read them, you must use a separate newsreader application and go to the site. Some browsers, such as Netscape or Internet Explorer, have newsreading capabilities. There are two major newsgroups for discussion of homebuying and home ownership:

1. misc.consumers.house
2. misc.invest.real-estate

CENTURY 21® Communities

If you're a subscriber to AOL, come visit us at CENTURY 21® Communities (keyword: CENTURY 21). You'll be able to search the site for information on hundreds of communities nationwide. Searching by zip code, you can zero in on cultural activities, lifestyle, recreation, and schools. Critical data on cost of living, jobs, and crime also are available. The site features live chats with relocation and homebuying experts.

You'll see excerpts from many of our CENTURY 21® series of groups. You'll get up-to-date information with text articles on a range of subjects. The site also will springboard you to a complete CENTURY 21® broker directory and broker Web sites.

Get Real!

The wealth of information online is astounding. Surfing is fun. To you virtual home hunters immersed in this online cornucopia, be sure to visit the physical world too. Walk the neighborhoods. Talk to agents. And happy home hunting from CENTURY 21®!

GLOSSARY

abstract of title (abstract) History of a parcel of real estate, compiled from public records, listing transfers of ownership and claims against the property

acceleration clause Provision in a mortgage document stating that if a payment is missed or any other provision violated, the whole debt becomes immediately due and payable

acknowledgment Formal declaration before a public official that one has signed a document

acre Land measure equal to 43,560 square feet

adjustable-rate mortgage (ARM) Loan whose interest rate is changed periodically to keep pace with current levels

adjusted basis Original cost of property plus any later improvements and minus a figure for depreciation claimed

adjusted sales price Sale price minus commissions, legal fees, and other costs of selling

agent Person authorized to act on behalf of another in dealing with third parties

agreement of sale (purchase agreement, sales agreement, contract to purchase) Written contract detailing terms under which buyer agrees to buy and seller agrees to sell

alienation clause (due-on-sale, nonassumption) Provision in a mortgage document stating that the loan must be paid in full if ownership is transferred, sometimes contingent upon other occurrences

amortization Gradual payment of a debt through regular installments that cover both principal and interest

appraisal Estimate of value of real estate

appreciation Increase in value or worth of property

"as is" Present condition of property being transferred, with no guaranty or warranty provided by the seller

assessed valuation Value placed on property as a basis for levying property taxes; not identical with appraised or market value

assignment Transfer of a contract from one party to another

assumable mortgage Loan that may be passed to the next owner of the property

assumption Takeover of a loan by any qualified buyer (available for FHA and VA loans)

automatic renewal clause Provision that allows a listing contract to be renewed indefinitely unless canceled by the property owner

balloon loan Mortgage in which the remaining balance becomes fully due and payable at a predetermined time

balloon payment Final payment on a balloon loan

bill of sale Written document transferring personal property

binder Preliminary agreement of sale, usually accompanied by earnest money (term also used with property insurance)

bond Roughly the same as *promissory note,* a written promise to repay a loan, often with an accompanying mortgage that pledges real estate as security

broker Person licensed by the state to represent another for a fee in real estate transactions

building code Regulations of local government stipulating requirements and standards for building and construction

buydown The payment of additional points to a mortgage lender in return for a lower interest rate on the loan

buyer's broker Agent who takes the buyer as client, is obligated to put the buyer's interests above all others, and owes specific fiduciary duties to the buyer

buyer's market Situation in which the supply of homes for sale exceeds the demand

cap Limit (typically about 2 percent) by which an adjustable mortgage rate might be increased at any one time

capital gain Taxable profit on the sale of an appreciated asset

caveat emptor Let the buyer beware

ceiling Also known as a *lifetime cap,* limit beyond which an adjustable mortgage rate may never be raised

certificate of occupancy Document issued by a local governmental agency stating that the property meets the standards for occupancy

chattel Personal property

client The broker's principal, to whom fiduciary duties are owed

closing (settlement, escrow, passing papers) Conclusion of a real estate sale, at which time title is transferred and necessary funds change hands

closing costs One-time charges paid by the buyer and the seller on the day the property changes hands

closing statement Statement prepared for the buyer and the seller listing debits and credits, completed by the person in charge of the closing

cloud (on title) Outstanding claim or encumbrance that challenges the owner's clear title

commission Fee paid (usually by a seller) for a broker's services in securing a buyer for property; commonly a percentage of sale price

commitment (letter) Written promise to grant a mortgage loan

common elements Parts of a condominium development in which each owner holds an interest (swimming pool, etc.)

comparable Recently sold similar property used to estimate the market value

comparative market analysis Method of valuing homes using the study of comparables, property that failed to sell, and other property that currently is on the market

conditional commitment Lender's promise to make a loan subject to the fulfillment of specified conditions

conditional offer Purchase offer in which the buyer proposes to purchase only after certain occurrences (sale of another home, securing of financing, etc.)

condominium Type of ownership involving individual ownership of dwelling units and common ownership of shared areas

consideration Anything of value given to induce another to enter into a contract

contingency Condition (inserted into the contract) that must be satisfied before the buyer purchases a house

contract Legally enforceable agreement to do (or not to do) a particular thing

contract for deed (land contract) Method of selling by which the buyer receives possession but the seller retains title

conventional mortgage Loan arranged between lender and borrower with no government guarantee or insurance

cost basis Accounting figure that includes original cost of property plus certain expenses to purchase and money spent on permanent improvements and other costs minus any depreciation claimed on tax returns over the years

curtesy In some states, rights a widower obtains to a portion of his deceased wife's real property

customer Typically the buyer as opposed to the principal (seller)

days on market (DOM) Number of days between the time a house is put on the market and the date of a firm sale contract

deed Formal written document transferring title to real estate; a new deed is used for each transfer

deed of trust Document by which title to property is held by a neutral third party until a debt is paid; used instead of a mortgage in some states

deed restriction (restrictive covenant) Provision placed in a deed to control the use and occupancy of the property by future owners

default Failure to make mortgage payment

deferred maintenance Needed repairs that have been put off

deficiency judgment Personal claim against the debtor when foreclosed property does not yield enough at sale to pay off loans against it

delivery Legal transfer of a deed to the new property owner; the moment at which transfer of title occurs

depreciation Decrease in value of property because of deterioration or obsolescence; sometimes, an artificial bookkeeping concept valuable as a tax shelter

direct endorsement Complete processing of an FHA mortgage application by an authorized local lender

documentary tax stamp Charge levied by state or local government when real estate is transferred or mortgaged

dower In some states, the rights of a widow to a portion of her deceased husband's property

down payment Cash to be paid by the buyer at closing

DVA Department of Veterans Affairs; formerly VA, Veterans Administration

earnest money Buyer's "good faith" deposit accompanying purchase offer

easement A permanent right to use another's property (telephone line, common driveway, footpath, etc.)

encroachment Unauthorized intrusion of a building or improvement onto another's land

encumbrance Claim against another's real estate (unpaid tax, mortgage, easement, etc.)

equity The money realized when property is sold and all the claims against it are paid; commonly, sale price minus present mortgage

escrow Funds given to a third party to be held pending some occurrence; may refer to earnest money, funds collected by a lender for the payment of taxes and insurance charges, funds withheld at closing to ensure uncompleted repairs, or in some states the entire process of closing

exclusive agency Listing agreement under which only the listing office can sell the property and keep the commission unless the owner sells the house, in which case no commission is paid

exclusive right to sell Listing agreement under which the owner promises to pay a commission if the property is sold during the listing period by anyone, even the owner

fair market value See **market value**

fee simple (absolute) Highest possible degree of ownership of land

FHA Federal Housing Administration (FHA), which insures mortgages to protect the lending institution in case of default

FHA mortgage Loan made by a local lending institution and insured by the FHA, with the borrower paying the premium

fiduciary A person in a position of trust or responsibility with specific duties to act in the best interest of the client

first mortgage Mortgage holding priority over the claims of subsequent lenders against the same property

fixture Personal property that has become part of the real estate

foreclosure Legal procedure for enforcing payment of a debt by seizing and selling the mortgaged property

front foot Measurement of land along a street or waterfront—each front foot is one foot wide and extends to the depth of the lot

grantee The buyer, who receives a deed

grantor The seller, who gives a deed

guaranteed sale Promise by the listing broker that if the property cannot be sold by a specific date, the broker will buy it, usually at a sharply discounted price

hazard insurance Insurance on a property against fire and similar risks

homeowners policy Policy that puts many kinds of insurance together into one package

improvements Permanent additions that increase the value of a home

index Benchmark measure of current interest levels used to calculate periodic changes in rates charged on adjustable-rate mortgages

joint tenancy Ownership by two or more persons, each with an undivided ownership—if one dies, the property goes automatically to the survivor

junior mortgage A mortgage subordinate to another

land contract Type of layaway installment plan for buying a house; sought by a buyer who does not have enough down payment to qualify for a bank loan or to persuade the seller to turn over title

lien A claim against property for the payment of a debt; mechanic's lien, mortgage, unpaid taxes, judgments

lis pendens Notice that litigation is pending on property

listing agreement (listing) Written employment agreement between a property owner and a real estate broker authorizing the broker to find a buyer

listing presentation Proposal submitted orally or in writing by a real estate agent who seeks to put a prospective seller's property on the market

loan servicing Handling paperwork of collecting loan payments, checking property tax and insurance coverage, and handling delinquencies

lock in To guarantee that the borrower will receive the rate in effect at the time of loan application

maintenance fees Payments made by the unit owner of a condominium to the homeowners association for expenses incurred in the upkeep of the common areas

margin Percentage (typically about 2.5 percent) added to the *index* to calculate the mortgage rate

marketable title Title free of *liens, clouds,* and *defects;* a title that will be freely accepted by a buyer

market value The most likely price a given property will bring if widely exposed on the market, assuming fully informed buyer and seller

mechanic's lien Claim placed against a property by unpaid workers or suppliers

meeting of the minds Agreement by a buyer and the seller on the provisions of a contract

mortgage A lien or claim against real property given as security for a loan; the homeowner "gives" the mortgage, the lender "takes" it

mortgagee The lender

mortgagor The borrower

Multiple Listing Service (MLS) Arrangement by which brokers work together on the sale of each other's listed homes with shared commissions

negative amortization Arrangement under which the shortfall in a mortgage payment is added to the amount borrowed; gradual raising of a debt

net listing Arrangement under which the seller receives a specific sum from the sale price and the agent keeps the rest as sales commission (open to abuses; illegal in most states)

note See **bond**

PITI Abbreviation for principal, interest, taxes, and insurance, often lumped together in a monthly mortgage payment

plat A map or chart of a lot, subdivision, or community showing boundary lines, buildings, and easements

PMI Private mortgage insurance

point (discount point) One percent of a new mortgage being placed; paid in a one-time lump sum to the lender

portfolio loans Loans made by a bank that keeps its mortgages as assets in its own portfolio (also called *nonconforming loans*)

prepayment Payment of a mortgage loan before its due date

prepayment penalty Charge levied by the lender for paying off a mortgage before its maturity date

principal The party (typically the seller) who hires and pays an agent

procuring cause Actions by a broker that bring about the desired results

prorations Expenses that are fairly divided between the buyer and the seller at closing

purchase-money mortgage Mortgage for the purchase of real property, commonly a mortgage "taken back" by the seller

quitclaim deed Deed that completely transfers whatever ownership the grantor may have had, but makes no claim of ownership in the first place

real property Land and the improvements on it

Realtist Member of the National Association of Real Estate Brokers

REALTOR® Registered name for a member of the National Association of REALTORS®

REALTOR-ASSOCIATE® Salesperson associated with a broker who is a member of a Board of REALTORS®

redlining The practice of refusing to provide loans or insurance in a certain neighborhood

RESPA Real Estate Settlement Procedures Act, requiring advance disclosure to the borrower of information pertinent to the loan

restrictive covenant See **deed restriction**

reverse mortgage Arrangement under which an elderly homeowner, who does not need to meet income or credit requirements, can draw against the equity in the home with no immediate repayment

salesperson Holder of an entry-level license who is allowed to assist a broker who is legally responsible for the salesperson's activities (synonymous in some areas with *agent*)

seller's broker Agent who takes the seller as a client, is legally obligated to a set of fiduciary duties, and is required to put the seller's interests above all others

seller's market Situation in which the demand for homes exceeds the supply offered for sale

settlement See **closing**

specific performance Lawsuit requesting that a contract be exactly carried out, usually asking that the seller be ordered to convey the property as previously agreed

subagency Legal process by which the seller who lists property for sale with a broker takes on the broker's associates and cooperating firms in a multiple listing system as agents

survey Map made by a licensed surveyor who measures the land and charts its boundaries, improvements, and relationship to the property surrounding it

time is of the essence Legal phrase in a contract requiring punctual performance of all obligations

title Rights of ownership, control, and possession of property

title insurance Policy protecting the insured against loss or damage due to defects in the title: The "owner's policy" protects the buyer, the "mortgagee's policy" protects the lender; paid with a one-time premium

title search Check of public records, usually at the local courthouse, to make sure that no adverse claims affect the value of the title

INDEX

ABOUT THE AUTHORS

The CENTURY 21® System, recognized as the number-one consumer brand in real estate, has helped millions of people with their homebuying and homeselling needs for more than 25 years. In the relocation process as well, the CENTURY 21® System has assisted families every step of the way.

To meet the high expectations of today's demanding, value-conscious consumer, the CENTURY 21® System has redefined the real estate industry with innovative technology tools, strategic alliances with other industry leaders that offer a wide array of home-oriented products and services, distinct brands for special properties, and other housing-related opportunities.

Visit the CENTURY 21® System at *Century 21 Communities*[SM], the most comprehensive source of real estate and community information on cities across North America, on America Online® at **Keyword: CENTURY 21.** Or contact one of the System's network of 6,400 independently owned and operated offices throughout the United States and Canada, as well as in 21 countries around the world.

Patrick Hogan is an editor and writer living in Chicago. He is coauthor of the Dearborn book *How to Avoid the 10 Biggest Homebuying Traps, 6th Edition,* and has edited numerous real estate, small business, and information technology books.

Special Offer!

You get all this with an
ADT Security System

- Up to 20% savings on homeowner's insurance
- An exclusive 6-month, money-back service guarantee
- A system that can meet your needs and budget
- 24-hour burglary and fire monitoring

Installation Savings That Can Really Add Up

Save **$50** on an installation cost of $249

Save **$75** on an installation cost of $499

Save **$100** on an installation cost of $999 or more

To take advantage of this special offer
call 1-800-221-6151!